CONCRETE & MASONRY

No. 902
$10.95

CONCRETE & MASONRY

This book, published by TAB BOOKS, Blue Ridge Summit, Pennsylvania, contains material from U.S. Government documents TM 5-742 and NAVTRA 10649-F.

First Edition
First Printing—March 1976

Hardbound Edition: International Standard Book No. 0-8306-6902-7

Paperbound Edition: International Standard Book No. 0-8306-5902-1

Library of Congress Card Number: 76-1553

Contents

Preface

Concrete and masonry construction were once the province of skilled journeymen who taught the trades to apprentices through on-the-job training programs. Times have changed. With labor costs soaring and material prices elevated to outlandish proportions, homeowners and would-be builders have in increasing numbers been turning to their local libraries for information that would allow them to bypass the traditional approaches to learning about construction and the requirements involved in application of the principles of building.

The information contained in this book is not the work of any one man—it is doubtful that a single author would possess the depth of knowledge required to prepare a work of this scope. Rather, it is the distillation of material prepared for the government under the auspices of the army's engineer school in Fort Belvoir, Virginia, incorporating selected passages from the navy's *Builder* rate training handbook and contributions from *Portland Cement Association*.

Concrete work, like masonry, is not particularly complicated. But there's more to turning out a professional, structurally reliable job than mixing some sand and concrete and pouring the mix. The proportions of mixes are vital to the integrity of your finished work; the water you use in your mix will affect the strength of the concrete; the temperature of the setting concrete is important, too. This book gives you *all* the information you need to understand the language of building and do the work yourself, from formulating mixtures in the proper ratios to constructing the wood forms that will contain the mixed concrete—from preparing the site to pouring and working the wet concrete. The final chapter in the section on concrete construction even tells you how and when to employ reinforcing elements, for construction of beams, columns, and slabs.

Masonry work is that part of construction that deals with the joining of concrete, brick, or tile blocks using mortar as the bonding agent. The text and illustrations in the section on masonry show clearly the right and wrong methods for constructing walls of brick, tile, or concrete block. Again, no prior knowledge on your part is assumed, and the field of masonry is covered beginning at ground zero. You'll learn the terms and the tools of the trade and then the actual construction techniques.

PART ONE
CONCRETE

1

Concrete as a Building Material

Advantages. Portland cement (para 2–2) is the most important masonry material used in modern construction. Its numerous advantages make it one of the most economical, versatile, and universally used construction materials available. It is commonly used for buildings, bridges, sewers, culverts, foundations, footings, piers, abutments, retaining walls, and pavements. A concrete structure, either plain or reinforced, is almost unique among the many systems of modern construction. In its plastic state concrete can be readily handled and placed in forms and cast into any desired shape. Quality concrete work produces structures which are lasting, pleasing in appearance, and require comparatively little maintenance.

Limitations. Recognition of the limitations of concrete construction in the design phase will eliminate some of the structural weaknesses that detract from the appearance and serviceability of concrete structures. Some of the principal limitations and disadvantages are:

Low tensile strength. Concrete members which are subjected to tensile stress must be reinforced with steel bars, high-strength steel wire, or mesh.

Drying shrinkage and moisture movements. Concrete, like all construction materials, contracts and expands under various conditions of moisture and/or temperature. This normal movement should be anticipated and provided for in the design, placement, and curing. Otherwise, damaging cracks may result.

Permeability. Even the best concrete is not entirely impervious to moisture. It contains soluble compounds which may be leached out to varying degrees by water. Impermeability is particularly important in reinforced concrete where reliance is placed on the concrete cover to prevent rusting of the steel, and where the structure is exposed to freezing and thawing.

Concrete Materials

Chemical Process. The essential ingredients of concrete are cement and water which react chemically in a process called hydration to form another material having useful strength. Hardening of concrete is not the result of the drying of the mix, as can be seen from the fact that fresh concrete placed under water will harden despite its completely submerged state. The mixture of cement and water is called cement paste, but such a mixture, in large quantities, is prohibitively expensive for practical construction purposes and undergoes excessive shrinkage upon hardening.

Aggregates. Inert filler materials in the form of sand, stone, and gravel are added to cement and water in prescribed amounts to increase the volume of the mixture. When concrete is properly mixed each particle of aggregate is completely surrounded by paste and all spaces between aggregate particles are completely filled. The paste is the cementing medium that binds the aggregate particles into a solid mass.

Grout. Grout is a mixture of portland cement, lime, fine aggregate, and water in such proportions that the mixture is fluid. Exact proportions and the maximum size of the aggregate are dictated by the intended purpose.

Mortar. Mortar is a mixture of portland cement, lime, fine aggregate, and water in such proportions that the mixture is plastic. Exact proportions and the maximum size of the aggregate are determined by the intended purpose.

DESIRABLE PROPERTIES OF CONCRETE

Plastic Concrete

A plastic concrete is a concrete mix that is readily molded, yet changes its shape slowly if the mold is immediately removed. The degree of plasticity influences the quality and character of the finished product. Control of the ingredients in the mix limits the variables to the proportions of the ingredients. Significant changes in the mix proportions are indicated by the slump (App. B). Desirable properties of the plastic concrete are:

Workability. This property indicates the relative ease or difficulty of placing and consolidating concrete in the form. The consistency of the mixture is measured by the slump test (app. B) and is maintained as necessary to obtain the required workability for the specific conditions and method of placement. A very stiff mix would have little slump and would be very difficult to place in heavily reinforced sections. It is a good mix to place in a slab where reinforcing is not used. A more fluid mix can be placed where reinforcing steel is present. Workability is controlled largely by the amounts of and proportion of fine to coarse aggregate used with a given quantity of paste.

Nonsegregation. A plastic concrete should be handled so that there will be a minimum of segregation and the mix will remain homogeneous. For example, to prevent segregation, plastic concrete should not be allowed to drop (free fall) more than 3 to 5 feet. Care must also be taken in handling to prevent bleeding.

Uniformity. For uniformity every batch should be accurately proportioned according to the specifications. Uniform quality of the hardened concrete is desirable from both economic and strength considerations.

Hardened Concrete

Hardened concrete in its finished form is the actual basis of any concrete design. The essential qualities which must be considered are:

Strength. Strength is the ability of the concrete to resist a load in compression, flexure, or shear. The principal influencing factor on strength is the ratio of water to cement. About 2½ gallons of water are required for hydration (chemical reaction with water) of a sack of cement. Additional water is used to thin the paste, thus allowing it to coat more particles. This increases the yield obtainable from each sack of cement, thereby producing a more economical mix. However, excessive water-cement ratios should be avoided because thin paste is weak and a reduction in strength of the hardened concrete occurs due to the dilution of the paste. The minimum and maximum amounts of water generally used for economical mixes range from 4 to 8 gallons per sack.

Durability. Durability in concrete is the ability of the hardened mass to resist the effects of the elements, such as the action of wind, frost, snow, and ice, the chemical reaction of soils, or the effects of salts and abrasion. Durability is affected by climate and exposure. As the water-cement ratio is increased, the durability will decrease correspondingly. Air-entraining cements produce concretes with improved durability.

Water-Tightness. Water-tightness is an essential requirement of concrete. Tests show that the water-tightness of the paste is dependent on the amount of mixing water that is used and the extent to which the chemical reactions between cement and water have progressed.

Concrete Components

Natural Cements

Natural cement is produced by the grinding and calcination of a natural cement rock which is a clayey limestone containing up to twenty-five percent of clayey material.

Characteristics of Natural Cement. Natural cement is normally yellow to brown in color. The tensile strength and compressive strength of natural cement mortars are low, varying from one-third to one-half of the strengths of normal portland cement. Natural cement is variable in quality and is little used today. In the United States, it represents no more than one percent of the production of all cements.

Uses of Natural Cement. Two types are available commercially: Type N natural cement and Type NA air-entraining natural cement. Natural cement is used in the preparation of masonry cements for use in mortar and in combination with portland cement for use in concrete mixtures.

Portland Cements

Description. Portland cements are mixtures of selected raw materials which are finely ground, proportioned and calcined to the fusion temperature (approx. 2700°F.) to give the desired chemical composition. The clinker resulting from calcination is then finely pulverized. When combined with water, these cements undergo a chemical reaction and harden to form a stone-like mass. This reaction is called hydration and these cements are termed hydraulic cements.

Manufacture. Raw materials used in the manufacture of portland cements may include limestone, cement rock, oyster shells, coquina shells, marl, clay, shale, silica sand, and iron ore. These materials are pulverized and mixed so that the appropriate proportions of lime, silica, alumina, and iron components are present in the final mixture. To accomplish this, either a dry or wet process is used. In the dry process, grinding and blending are done with dry materials. In the wet process, the grinding and blending operations utilize a watery slurry. The prepared mixture is then fed into a rotary kiln which produces temperatures of 2600° to 3000°F. During this process, several reactions occur which result in the formation of portland cement clinker. The clinker is cooled and then pulverized with a small amount of gypsum added to regulate the setting time. The pulverized product is the finished portland cement. It is ground so fine that nearly all of it will pass through a sieve having 200 meshes to the lineal inch or 40,000 openings in a square inch. Each manufacturer of portland cement uses a trade or brand name under which the product is sold.

Common ASTM Portland Cements. Different types of portland cements are manufactured to meet certain physical and chemical requirements for specific puroses. The American Society for Testing and Materials (ASTM) provides for five types of portland cement in ASTM C150, "Standard Specifications for Portland Cement".

ASTM Type I. This type is also called normal portland cement and is a general-purpose cement suitable for all uses when the special properties of the other types are not required. Type I portland cement is more generally available than are the other types of cement. It is used in pavement and sidewalk construction, reinforced concrete buildings and bridges, railways, tanks, reservoirs, sewers, culverts, water-pipes, masonry units, and soil cement mixtures. In general, it is used when concrete is not subject to special sulfate hazard or where the heat generated by the hydration of the cement will not cause an objectionable rise in temperature.

ASTM Type II. This type is a modified cement used where precaution against moderate sulfate attack is important, as in drainage structures where the sulfate concentrations in the soil or ground water are higher than normal, but not unusually severe. Type II will usually generate less heat at a slower rate than Type I. It may be used in structures of considerable size where cement of moderate heat of hydration will tend to minimize temperature rise, as in large piers, heavy abutments, and heavy retaining walls and when the concrete is placed in warm weather.

ASTM Type III. Type III is a high-early-strength cement which provides high strengths at an early period, usually a week or less. Concrete made with Type III cement has a 7-day strength comparable to the 28-day strength of concrete made with Type I cement, and a 3-day strength comparable to the 7-day strength of concrete made with Type I cement. Type III cement has a higher heat of hydration and is more finely ground than

Type I cement. It is used where it is desired to remove forms as soon as possible, to put the concrete in service as quickly as possible, and in cold weather construction to reduce the period of protection against low temperatures. Although richer mixtures of Type I may be used to gain high-early-strength, Type III may provide it more satisfactorily and/or more economically.

ASTM Type IV. Type IV is a low heat cement for use where the rate and amount of heat generated must be minimized. The development of strength is also at a slower rate. It is intended for use only in large masses of concrete such as large gravity dams where temperature rise resulting from the heat generated during hardening is a critical factor.

ASTM Type V. Type V is a sulfate-resistant cement used only in concrete exposed to severe sulfate action. It is used principally where soil or groundwater in contact with the concrete structure has a high sulfate content. It gains strength more slowly than Type I.

Other ASTM Portland Cements. The American Society for Testing Materials covers certain other types of portland cements in separate ASTM specifications. These types are:

Air-entraining portland cements. Specifications for three types—Types IA, IIA, and IIIA—are given in AST MC175. They correspond in composition to Types I, II, and III, respectively in ASTM 150 with the addition of small quantities of air-entraining materials interground with the clinker during manufacture. These cements produce concrete with improved resistance to freeze-thaw action and to scaling caused by chemicals applied for snow and ice removal. Such concrete contains minute, well-distributed, and completely separated air bubbles.

White portland cement. White portland cement conforms to the specifications of ASTM C150 and C175. The finished product is white instead of gray. It is used primarily for architectual purposes.

Portland blast-furnace slag cements. The cements include two types conforming to the requirements of ASTM C595—Type IS and Type IS–A. To the latter, air-entraining additive has been added. In producing these cements, granulated blast furnace slag is either interground with portland cement clinker or blended with portland cement. These cements can be used in general concrete construction.

Portland-pozzolan cements. Portland-pozzolan cements include four types (P, IP, P–A, and IP–A, the latter two containing an air-entraining additive) as specified in ASTM C595. In these cements, pozzolan consisting of siliceous, or siliceous and aluminous material is blended with ground portland cement clinker. They are used principally for large hydraulic structures such as bridge piers and dams. The comparative strength of concrete made with portland-pozzolan cements may be lower than that made with normal cements.

Masonry cements. Masonry cements are mixtures of portland cement, air-entraining additives, and supplemental materials selected for their ability to impart workability, plasticity, and water retention to masonry mortars. These cements conform to the requirements of ASTM C91.

Special Portland Cements. In addition to the above cements, there are special types of portland cement not covered by ASTM specifications.

Oil well cement. Oil well portland cement is made to harden properly at the high temperatures prevailing in very deep oil wells.

Waterproofed portland cement. Waterproofed portland cement is made by grinding water-repellant materials with the clinker from which it is made.

Plastic cements. Plastic cements are made by adding plasticizing agents to the clinker. It is commonly used for making mortar, plaster, and stucco.

Packaging and Shipping. Cement is shipped either in sacks which weigh 94 pounds and have a loose volume of 1 cubic foot or in bulk by railroad, truck, or barge. Cement requirements for large projects may be given in terms of barrels, a barrel being equivalent to 4 sacks or 376 pounds.

Storage. Portland cement that is kept dry retains its quality indefinitely. Cement which has been stored in contact with moisture sets more slowly and has less strength than dry cement. Sacked cement should be stored in a warehouse or shed as nearly airtight as possible. The sacks should be stored close together (to reduce the circulation of air) and away from outside walls. Bags to be stored for long periods should be covered with tarpaulins or other waterproof covering. If no shed is available, the sacks should be placed on raised wooden platforms. Waterproof coverings should be placed over the pile in such a way that rain cannot reach the cement or the platform. Occasionally, sacked cement in storage will develop what is commonly called "warehouse pack", a condition resulting from packing too tightly. The cement retains its quality under these conditions and the condition can usually be corrected by rolling the sacks on the floor. Cement should be freeflowing and free of lumps. If lumps exist that are hard to break up, the cement should be tested to determine its suitability. Hard lumps indicate partial hydration and will reduce the strength and durability of the concrete. Bulk cement is usually stored in weatherproof bins. Ordinarily, it does not remain in storage very long but it can be stored for a relatively long time without deterioration.

WATER

The purpose of water in the concrete mix is to combine with the cement in the hydration process, coat the aggregate, and permit the mix to be worked.

Impurities

Mixing water should be clean, free from organic materials, alkalies, acids, and oil. In general, water that is fit to drink is suitable for mixing with cement. However, water with excessive quantities of sulfates should be avoided even though it may be fit to drink. Otherwise, the result is a weak paste that may contribute to deterioration or failure of the concrete. Water of unknown quality may be used for making concrete if mortar cubes made with this water have 7- and 28-day strengths equal to at least 90 percent of companion specimens made with drinkable water. Tests should also be made to be sure that the setting time of the cement is not adversely affected by impurities in mixing water. Impurities in mixing water, when excessive, may affect not only setting time, concrete strength, and volume constancy, but may cause efflorescence or corrosion of reinforcement. In some cases, it may be necessary to increase the cement content of the concrete to compensate for the impurities. The effects of certain common impurities in mixing water on the quality of plain concrete are given below:

Alkali Carbonate and Bicarbonate. Carbonates and bicarbonates of sodium and potassium may either accelerate or retard the set of different cements. In large concentrations these salts can materially reduce concrete strength. Tests should be made when the sum of these dissolved salts exceeds 1,000 ppm (parts per million).

Sodium Chloride and Sodium Sulfate. A high dissolved solids content of a natural water is usually the result of a high content of sodium chloride or sodium sulfate. Concentrations of 20,000 ppm of sodium chloride and 10,000 ppm of sodium sulfate are generally tolerable.

Other Common Salts. Carbonates of calcium and magnesium are seldom found in sufficient concentration to affect the strength of concrete. Bicarbonates of calicum and magnesium may be present in concentrations up to 400 ppm without adverse effect. Magnesium sulfate and magnesium chloride can be present in concentrations up to 40,000 ppm without harmful effects on strength. Calcium chloride may be used to accelerate both hardening and strength gain.

Iron Salts. Natural groundwater usually contains only small quantities of iron. However, acid mine waters may contain large quantities of iron. Iron salts in concentrations up to 40,000 ppm can be tolerated.

Miscellaneous Inorganic Salts. Salts of manganese, tin, zinc, copper, and lead may cause a significant reduction in strength and cause large variations in setting time. Sodium iodate, sodium phosphate, sodium arsenate, and sodium borate can greatly retard the set and the strength development. Concentrations of these salts up to 500 ppm are acceptable in mixing water. Concentrations of sodium sulfide of even 100 ppm may be harmful.

Seawater. Seawater containing up to 35,000 ppm of salt is generally suitable for unreinforced concrete. Some strength reduction occurs but this may be allowed for by reducing the water-cement ratio. The use of seawater in reinforced concrete may increase the risk of corrosion of reinforcing steel; however, this risk is reduced if the reinforcement has sufficient cover and if the concrete is watertight and contains an adequate amount of entrained air. Seawater should not be used for making prestressed concrete in which the prestressing steel is in contact with the concrete.

Acid Waters. If possible, the acceptance of acid mixing water should be based on the concentration (in parts per million) of acids in the water rather than the pH of the water. Hydrochloric, sulfuric, and other common inorganic acids in concentrations up to 10,000 ppm generally have no adverse effect on concrete strength.

Alkaline Waters. Concentrations of sodium hydroxide above .5 percent by weight of cement may reduce the concrete strength. Potassium hydroxide in concentrations up to 1.2 percent by weight of cement has little effect on the concrete strength developed by some cements but substantially reduces the strength of other cements. If in doubt, tests should be made.

Industrial Waste Waters. Industrial waste waters carrying less than 4,000 ppm of total solids generally produce a reduction in compressive strength of no more than 10 percent. Water that contains unusual solids, such as those from tanneries, paint factories, coke plants, chemical and galvanizing plants, and so on, should be tested.

Waters Carrying Sanitary Sewage. Sewage which has been diluted in a good disposal system generally has no significant effect on the concrete strength.

Sugar. Small amounts of sugar, .03 to .15 percent weight of cement, usually retard the setting of cement. When the amount is increased to about .20 percent weight of cement, the set is usually accelerated. Sugar in quantities about .25 percent by weight of cement may cause substantial reduction in strength. If the concentration of sugar in the mixing water exceeds 500 ppm, tests should be made.

Silt or Suspended Particles. Mixing water may contain up to 2,000 ppm of suspended clay or fine rock particles without adverse effect.

Oils. Mineral oil, not mixed with animal or vegetable oils, probably has less effect on strength development than other oils. However, mineral oil in concentrations greater than 2 percent by weight of concrete may reduce the concrete strength by more than 20 percent.

Algae. Mixing water containing algae may result in excessive reduction in concrete strength. Algae may also be present on aggregates, in which case the bond between the aggregate and the cement paste is weakened.

AGGREGATES

Even though aggregates are considered as inert materials acting as filler, they make up from 60 to 80 percent of the volume of concrete. The characteristics of the aggregates have a considerable influence on the mix proportions and on the economy of the concrete. For example, rough-textured or flat and elongated particles require more water to produce workable concrete than do rounded or cubical particles. Hence, aggregate particles that are angular require more cement to maintain the same water-cement ratio and thus the concrete is more expensive. For most purposes, aggregates should consist of clean, hard, strong, durable particles free of chemicals or coatings of clay or other fine materials that affect the bond of the cement.

Characteristics of Aggregates

Contaminating materials most often encountered are dirt, silt, clay, mica, salts, and also humus or other organic matter that may appear as a coating or as loose, fine material. Many of them can be removed by washing but weak, friable or laminated aggregate particles are undesirable. Sand containing organic material cannot be washed clear. Shale or stones with shale and most cherts are especially undesirable. Visual inspection often discloses weaknesses in course aggregate. It should be tested in doubtful cases. The most commonly used aggregates are sand, gravel, crushed stone, and blast furnace slag. Cinders, burnt clay, expanded blast furnace slag, and other materials are also used. These aggregates produce normal-weight concrete, that is, concrete weighing from about 135 to 160 pounds per cubic foot. Normal-weight aggregates should meet the requirements of Specifications for Concrete Aggregates (ASTM C33). These specifications limit the permissible amounts of deleterious substances and cover requirements for gradation, abrasion resistance, and soundness. Aggregate characteristics, their significance, and standard tests for evaluation of these characteristics are given in table 2–1.

Abrasion Resistance. A general index of aggregate quality is the abrasion resistance of the aggregate. This characteristic is essential when the aggregate is used in concrete subject to abrasion as in heavy-duty floors.

Resistance to Freezing and Thawing. The freeze-thaw resistance of an aggregate is related to its porsoity, absorption, and pore structure. This is an important characteristic in exposed concrete. If an aggregate particle absorbs so much water that insufficient pore space is available, it will not accommodate water expansion that occurs during freezing. The performance of aggregates under exposure to freezing and thawing can be predicted in two ways: past performance and freezing-thawing tests of concrete specimens. If aggregates from the same source have previously given satisfactory service when used in concrete, the aggregate may be considered suitable. Aggregates not having a service record may be considered acceptable if they perform satisfactorily in concrete specimens subjected to freezing thawing tests (ASTM C290 and C291) and strength tests.

Chemical Stability. Aggregates which have chemical stability will neither react chemically with cement in a harmful manner nor be affected chemically by other external influences. Field service records generally provide the best information for the selection of nonreactive aggregates. If an aggregate has no service record and is suspected of being chemically unsound, laboratory tests are useful for determining its suitability. There are three ASTM tests for identifying alkali-reactive aggregates (ASTM C227, C289, and C586). In addition there is an ASTM recommended practice (C295) for testing of aggregates based on classification of the rock samples.

Particle Shape and Surface Texture. The particle shape and surface texture of an aggregate

Table 2–1. Characteristics of Aggregates

Characteristic	Significance or importance	Test or practice ASTM designation	Specification requirement
RESISTANCE TO ABRASION	Index of aggregate quality. Warehouse floors, loading platforms, pavements.	C131	Max. percent loss*
RESISTANCE TO FREEZING AND THAWING	Structures subjected to weathering.	C290.C291	Max. number of cycles
CHEMICAL STABILITY	Strength and durability of all types of structures	C227 (mortar bar) C289 (chemical) C586 (aggregate prism) C295 (petrographic)	Max. expansion of mortar bar* Aggregates must not be reactive with cement alkalies*
PARTICLE SHAPE AND SURFACE TEXTURE	Workability of fresh concrete		Max. percent flat and elongated pieces
GRADING	Workability of fresh concrete. Economy	C136	Max. and min. percent passing standard sieves
BULK UNIT WEIGHT	Mix design calculations. Classification	C29	Max. or min. unit weight (special concrete)
SPECIFIC GRAVITY	Mix design calculations.	C127 (coarse aggregate) C128 (fine aggregate)	
Absorption and surface moisture	Control of concrete quality.	C70, C127, C128	

*Aggregates not conforming to specification requirements may be used if service records or performance tests indicate they produce concrete having the desired properties.

influence the properties of fresh concrete more than they affect the properties of hardened concrete. Very sharp and rough aggregate particles or flat, enlongated particles require more fine material to produce workable concrete than do aggregate particles that are more rounded or cubical. Stones which break up into long slivery pieces should be avoided or limited to about 15 percent in either fine or coarse aggregate.

Grading. The grading and maximum size of aggregate are important because of their relative effect on the workability, economy, porosity, and shrinkage of the concrete. Experience shows that either very fine or very coarse sands are objectionable; the first is uneconomical, the last gives harsh, unworkable mixes. The gradation or particle-size distribution of aggregate is determined by a sieve analysis. The standard sieves used for this purpose are numbers 4, 8, 16, 30, 50, and 100 for fine aggregate. The standard sieves used for the determination of the fineness modulus for coarse aggregate are 6-, 3-, 1½-, ¾-, and ⅜-inch and number 4. These sieves, for both fine and coarse aggregate grading, are based on square openings, the size of the openings in consecutive sieves being related by a constant ratio. Other sieves may be used for coarse aggregate grading, however. Grading charts, convenient for showing size distribution, generally have lines representing successive standard sieves placed at equal intervals as shown in figure 2–1. This figure also shows the grading limits for fine aggregates and for one designated size of coarse aggregate (ASTM C33). Fine aggregate is material which will pass a number 4 sieve and be retained on a number 100 sieve. Coarse aggregate is material retained by a number 4 sieve.

Fineness modulus. This is a term used as an index to the fineness or coarseness of aggregate. It is the summation of the cumulative percentages of the material retained on the standard sieves divided by 100. It is not an indication of grading, for an infinite number of gradings will give the same value for fineness modulus. To obtain the fineness modulus, quarter a sample of at least 500 grams of sand, and sieve through the No. 4, 8, 16, 30, 50, and 100 sieves. Weigh the material retained on each sieve and then calculate the cumulative weights retained. Determine the fineness modulus by adding up the cumulative percents and dividing by 100 (fig. 2–2). In general, fine aggregate with a very high or a very low value for fineness modulus will not be as good for concrete aggregate as medium sand. Coarse sand may not be workable, and fine sands are uneconomical. The actual graduation of material is plotted on DD Form 1207 (Aggregate Grading Chart) to determine if the sand is within specifications. Care should be taken to obtain representative samples. The aggregate fineness modulus from one

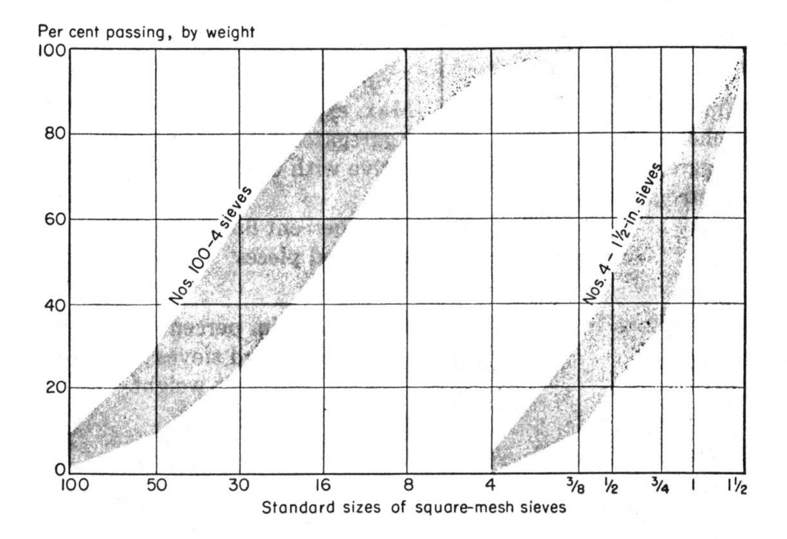

Figure 2–1. Limits specified in ASTM C33 for fine aggregates and for one size of coarse aggregate.

SCREEN SIZE	WEIGHT RETAINED (GRAMS) INDIVIDUAL	CUMULATIVE	CUMULATIVE % RETAINED
NO. 4	40	40	40
NO. 8	130	170	17.0
NO. 16	130	300	30.0
NO. 30	250	550	55.0
NO. 50	270	820	82.0
NO. 100	100	920	92.0
PAN	80	--	--
TOTAL WEIGHT	1000	--	280.0
FINENESS MODULUS (FM) = $\frac{280}{100}$ = 2.80			

Figure 2–2. Typical calculation of fineness modulus.

source should not vary more than 0.20 from test samples taken at the source.

Fine-aggregate grading. The most desirable fine aggregate grading depends on the type of work, richness of mix, and maximum size of coarse aggregate. In leaner mixes, or when small-size coarse aggregates are used, a grading that approaches the maximum recommended percentage passing each sieve is desirable for workability. In richer mixes, coarser gradings are desirable for economy. In general, if the water-cement ratio is kept constant and the ratio of fine to coarse aggregate is chosen correctly, a wide range in grading can be used without measurable effect on strength. The amount of fine aggregate passing the No. 50 and 100 sieves affects workability, finish and surface texture and water gain. For thin walls, hard-finished concrete floors, and smooth surfaces where concrete is cast against forms, the fine aggregate should contain not less than 15 percent passing the No. 50 sieve and at least 3 or 4 percent but not more than 10 percent passing the No. 100 sieve. With these minimum amounts of fines the concrete has better workability and is more cohesive so there is less water gain or bleeding than when lower percentages of fines are present. Aggregate gradings within the limits of ASTM C33 are generally satisfactory for most concretes.

Coarse-aggregate grading. The grading of a coarse aggregate of a given maximum size may be varied over a relatively wide range without appreciable effect on the cement and water requirements if the proportion of fine aggregate produces concrete of good workability. Table 2–2 indicates the gradation requirements for coarse aggregate. If wide variations occur in coarse aggregate grading, it may be necessary to vary the mix proportions in order to produce workable concrete. In such cases, it is often more economical to maintain grading uniformity in handling and manufacturing coarse aggregate than to adjust proportions for variations in gradation. Coarse aggregate should be graded up to the largest size that is practicable to use for the conditions of the job. The maximum size should not exceed ⅕ the dimension of nonreinforced members, ¾ of the clear spacing between reinforcing bars or between reinforcing bars and forms, and ⅓ the depth of nonreinforced slabs on ground. The larger the maximum size of the coarse aggregate, the less will be the mortar and paste necessary and the less water and cement required to produce a given quality. Field experience indicates that the amount of water required per unit volume of concrete for a given consistency and given aggregates is substantially constant regardless of the cement content or relative proportions of water to cement. Further, the water required decreases with in-

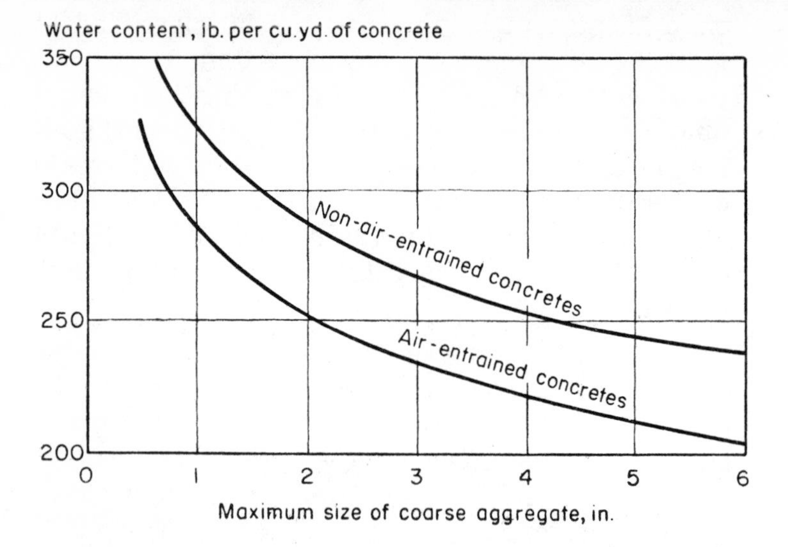

Figure 2–3. Water requirement for concrete of a given consistency as a function of coarse aggregate size.

creases in the maximum size of the aggregate. The water required per cubic yard of concrete with a slump of 3 to 4 inches is shown in figure 2–3 for a wide range in coarse aggregate sizes. It is apparent that, for a given water-cement ratio, the amount of cement required decreases (consequently, economy increases) as the maximum size of coarse aggregate increases. However, in some instances, especially in higher strength ranges, concrete with smaller maximum-size aggregate has a higher compressive strength than concrete with larger maximum-size aggregate at the same water-cement ratio.

Gap-graded aggregates. Certain particle sizes are lacking in gap-graded aggregates. Lack of two or more successive sizes may result in segregation problems, especially for non-air-entrained concretes with slumps greater than about 3 inches. If a stiff mix is required, gap-graded aggregates may produce higher strengths than normal aggregates used with comparable cement contents.

Bulk Unit Weight. The bulk unit weight of an aggregate is the weight of the material used to fill a one-cubic-foot container. The term "bulk unit weight" is used since the volume contains both aggregate and voids. Methods of determining bulk unit weights of aggregates are given in ASTM C29.

Specific Gravity. The specific gravity of an aggregate is the ratio of its weight to the weight of an equal volume of water. Most normal-weight aggregates have specific gravities of 2.4 to 2.9. Methods of determining specific gravity for coarse and fine aggregates are given in ASTM C127. In concrete calculations, the specific gravities used are generally given for saturated, surface-dry aggregates; that is, all pores are filled with water, but no excess moisture is present on the surface. The internal structure of an aggregate particle is made up of solid matter and voids that may or may not contain water.

Absorption and Surface Moisture. Is is necessary to determine the absorption and surface moisture of aggregates so that the net water content of the concrete can be controlled and correct batch weights determined. The moisture conditions of aggregates are depicted in figure 2–4. The four states are—

(1) *Oven-dry*—pores bone-dry, fully absorbent.

(2) *Air-dry*—Dry at the surface but containing some interior moisture, thus somewhat absorbent.

(3) *Saturated surface-dry*—neither absorbing water from nor contributing water to the concrete mix.

*Table 2-2. Gradation Requirements for Coarse Aggregate**

Amount finer than each laboratory sieve (square openings), percent by weight

Size Number	Nominal size (sieves) with square openings	4 in.	3½ in.	3 in.	2½ in.	2 in.	1½ in.	1 in.
1	3½ to 1½ in.	100	90 to 100		25 to 60		0 to 15	
2	2½ to 1½ in.			100	90 to 100	35 to 70	0 to 15	
357	2 in. to No. 4				100	95 to 100		35 to 70
467	1½ in. to No. 4					100	95 to 100	
57	1 in. to No. 4						100	95 to 100
67	¾ in. to No. 4							100
7	½ in. to No. 4							
8	⅜ in. to No. 8							
3	2 to 1 in.				100	90 to 100	35 to 70	0 to 15
4	1½ to ¾ in.					100	90 to 100	20 to 55

Size Number	Nominal size (sieves) with square openings	¾ in.	½ in.	⅜ in.	No. 4 (4760-micron)	No. 8 (2380-micron)	No. 16 (1190-micron)
1	3½ to 1½ in.	0 to 5					
2	2½ to 1½ in.	0 to 5					
357	2 in. to No. 4		10 to 30		0 to 5		
467	1½ in. to No. 4	35 to 70		10 to 30	0 to 5		
57	1 in. to No. 4		25 to 100		0 to 10	0 to 5	
67	¾ in. to No. 4	90 to 100		20 to 55	0 to 10	0 to 5	
7	½ in. to No. 4	100	90 to 100	40 to 70	0 to 15	0 to 5	
8	⅜ in. to No. 8		100	85 to 100	10 to 30	0 to 10	0 to 5
3	2 to 1 in.		0 to 5				
4	1½ to ¾ in.	0 to 15		0 to 5			

*From specifications for concrete aggregate (ASTM-C33).

(4) *Damp or wet*—containing an excess of moisture on the surface. When fine aggregate is damp and is handled, bulking generally occurs. Bulking is the increase in volume caused by surface moisture holding the particles apart. The variation in the amount of bulking with the moisture content and grading is shown in figure 2–5. Since most sands are delivered in a damp condition, wide variations can occur in batch quantities if the batching is done by volume. Such variations are likely to be out of proportion to the moisture content of the sand. For this reason, proportioning by volume, if undertaken, should be done with considerable care.

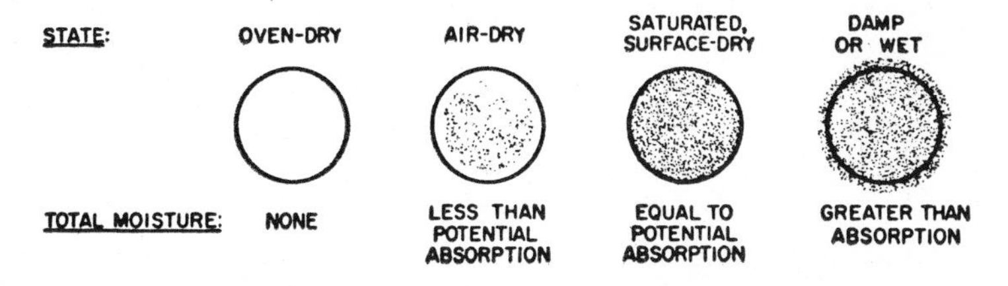

Figure 2–4. Moisture conditions of aggregates.

Deleterious Substances in Aggregates

Deleterious substances that may be present in aggregates include organic impurities, silt, clay, coal, lignite, and certain lightweight and soft particles. The effects of these substances on concrete and the ASTM test method designations are summarized in table 2–3.

Handling and Storing Aggregates

Aggregates should be handled and stored in such a manner as to minimize segregation and prevent contamination with deleterious substances. Aggregate is normally stored in stockpiles built up in layers of uniform thickness. Stockpiles should not be built up in high cone shapes nor allowed to run

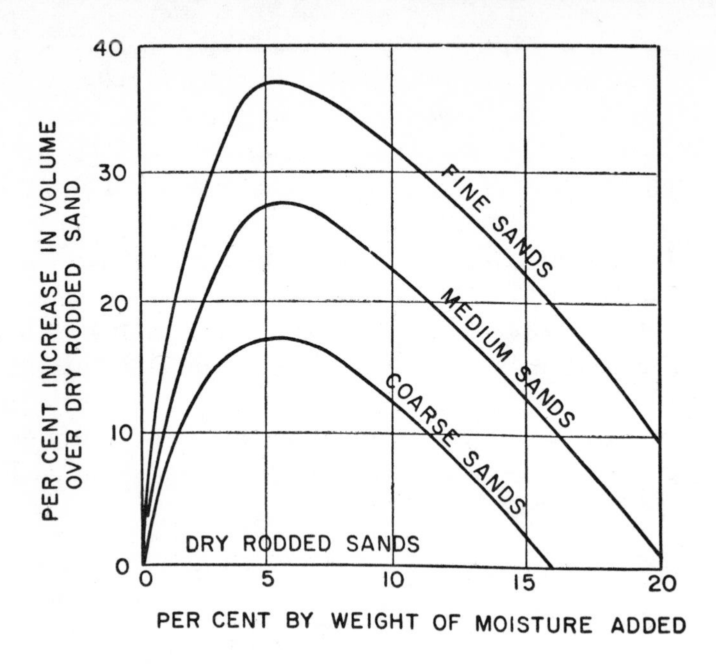

2–5. Variation in bulking with moisture and aggregate grading.

Table 2–3. Deleterious Substances in Aggregates

Deleterious substances	Effect on concrete	Test ASTM designation
ORGANIC IMPURITIES	Affect setting and hardening, and may cause deterioration	C40 C87
MATERIALS FINER THAN NO. 200 SIEVE	Affect bond, and increase water requirement	C–117
COAL, LIGNITE, OR OTHER LIGHTWEIGHT MATERIALS	Affect durability, and may cause stains and popouts	C123
SOFT PARTICLES	Affect durability	C235
FRIABLE PARTICLES	Affect workability and durability, and may cause popouts	C142

down slopes because this causes segregation. Aggregate should not be allowed to fall freely from the end of a conveyor belt. To minimize segregation, materials should be removed from stockpiles in approximately horizontal layers. If batching equipment is used, some of the aggregate will be stored in bins. Bins should be loaded by allowing the material to fall vertically over the outlet. Chuting the material at an angle against the side of the bin causes segregation of particles. Correct and incorrect methods of handling and storing aggregate are shown in figure 2–6.

ADMIXTURES

Definition and Purpose

Admixtures include all materials other than portland cement, water, and aggregates that are added to concrete, mortar, or grout immediately before or during mixing. Admixtures are sometimes used in concrete mixtures to improve certain quantities such as workability, strength, durability, watertightness, and wear resistance. They may also be added to reduce segregation, reduce heat of hydration, entrain air, and accelerate or retard setting and hardening. The same results can often be obtained by changing the mix proportions or by selecting other suitable materials without resort to admixtures (except air-entraining admixtures when necessary). Whenever possible, comparison should be made between these alternatives to determine which is more economical and/or convenient.

Air-Entrained Concrete

One of the major advances in concrete technology in recent years has been the advent of air entrainment. The use of entrained air is recommended in

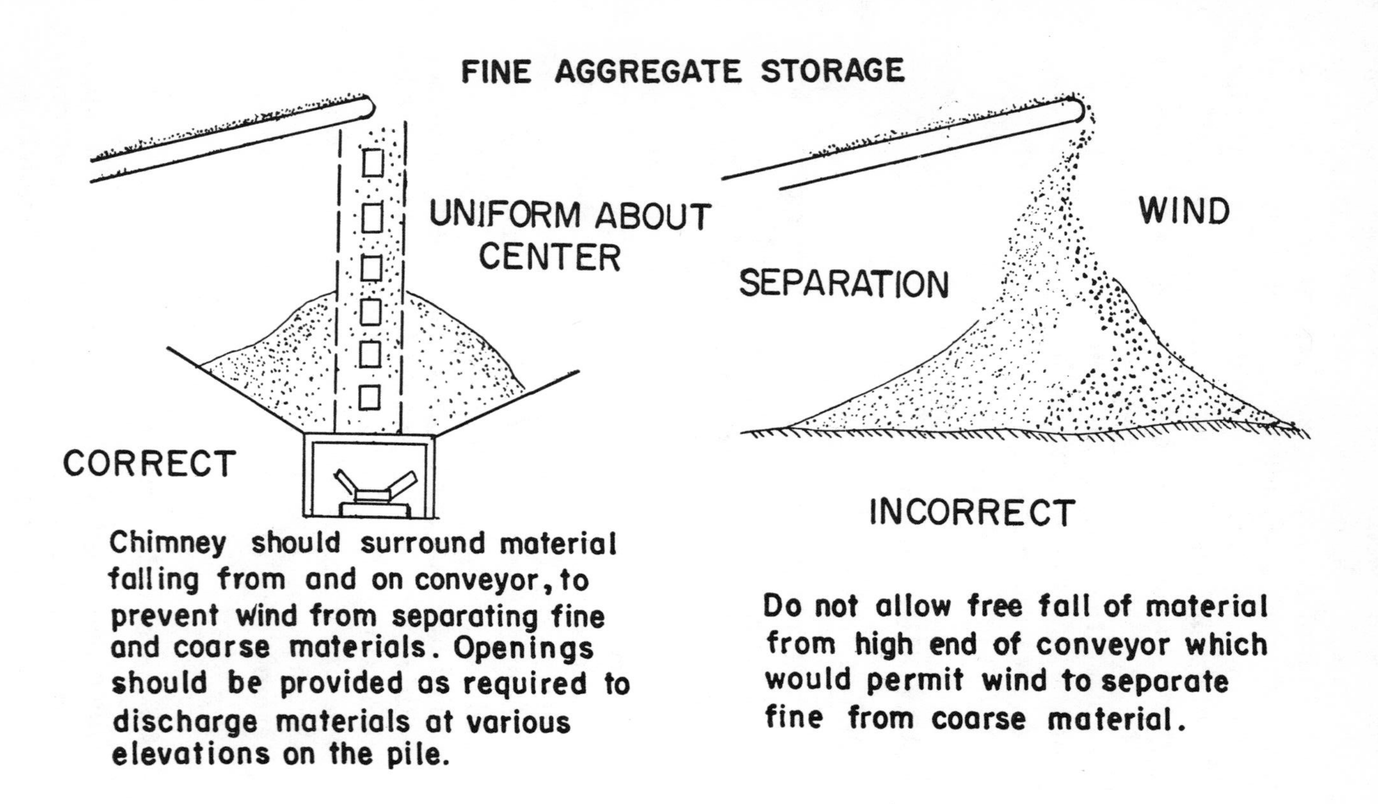
FINE AGGREGATE STORAGE
UNIFORM ABOUT CENTER
CORRECT
SEPARATION
WIND
INCORRECT
Chimney should surround material falling from and on conveyor, to prevent wind from separating fine and coarse materials. Openings should be provided as required to discharge materials at various elevations on the pile.
Do not allow free fall of material from high end of conveyor which would permit wind to separate fine from coarse material.

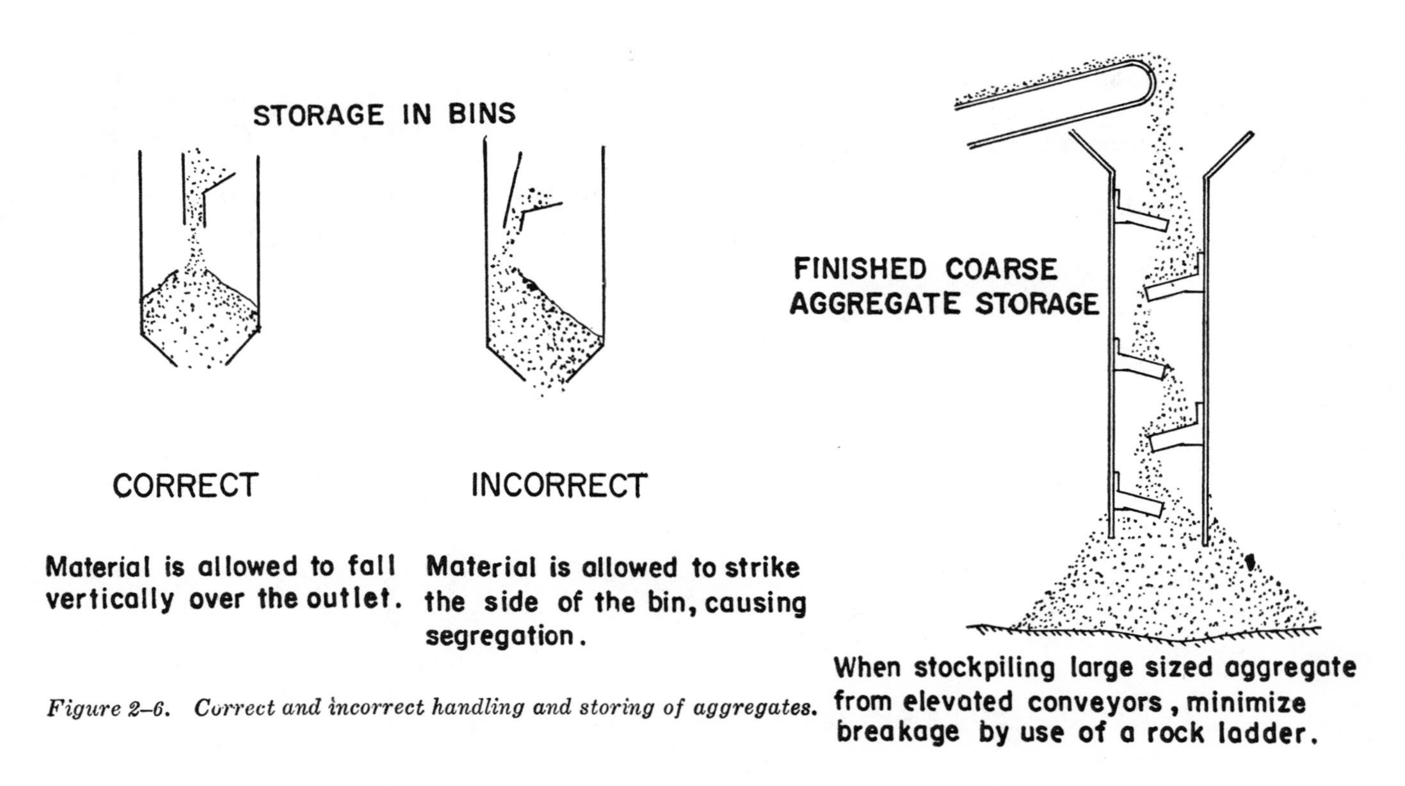

Figure 2–6. Correct and incorrect handling and storing of aggregates.

STOCKPILING OF COARSE AGGREGATE

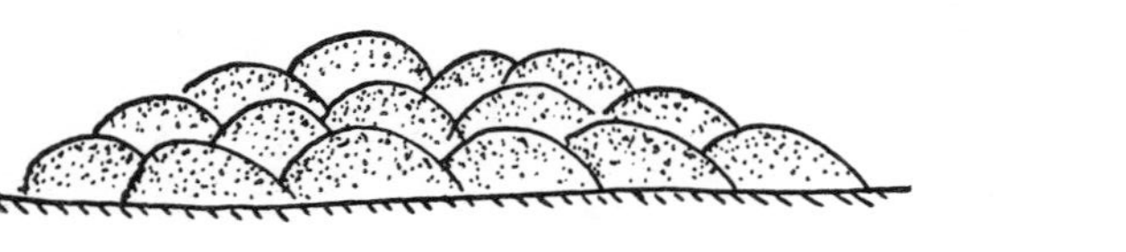

PREFERABLE

A crane or other equipement should stockpile material in separate batches, each no larger than a truckload, so that it remains where placed and does not run down slopes.

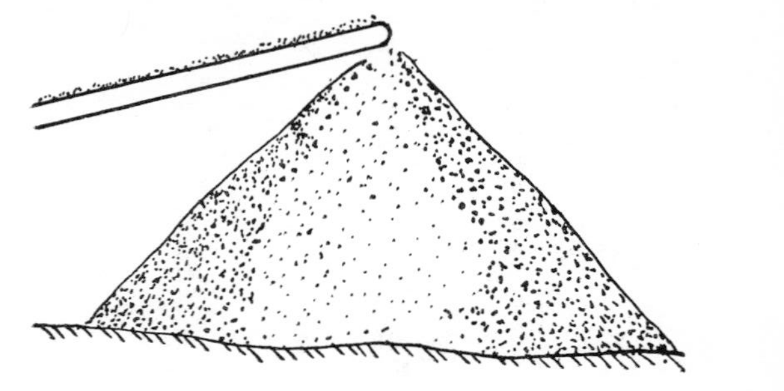

OBJECTIONABLE

Do not use methods that permit the aggregate to roll down the slope as it is added to the pile, or permit hauling equipment to operate over the same level repeatedly.

LIMITED ACCEPTABILITY

Generally, a pile should not be built radially in horizontal layers by a bulldozer working with materials as dropped from a conveyor belt. A rock ladder may be needed in this setup

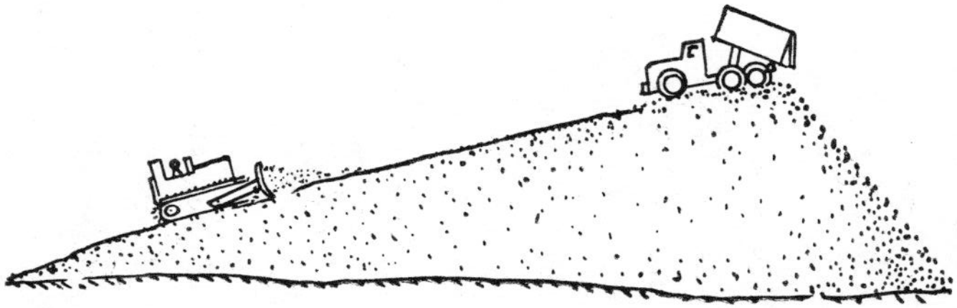

GENERALLY OBJECTIONABLE

A bulldozer stacking progressive layers on a slope not flatter than 3:1 is often objectionable unless materials strongly resist breakage.

Figure 2-6.—Continued.

concrete for most purposes. The principal reason for using intentionally entrained air is to improve concrete's resistance to freezing and thawing exposure. However, there are other important beneficial effects in both freshly mixed and hardened concrete. Air-entrained concrete is produced by using either an air-entraining cement or an air-entraining admixture during the mixing of the concrete. Unlike air entrapped in non-air-entrained concrete which exists in the form of relatively large air voids, which are not dispersed uniformly throughout the mix, entrained air exists in the form of minute disconnected bubbles well dispersed throughout the mass. These bubbles have diameters ranging from about one to three thousandths of an inch. As shown in figure 2–7, the bubbles are not interconnected and are well distributed throughout the paste. In general, air-entraining agents are derivatives of natural wood resins, animal or vegetable fats or oils, alkali salts of sulfated or sulfonated organic compounds and water-soluble soaps. Most air-entraining agents are in liquid form for use in the mix water. Instructions for the use of the various agents to produce a specified air content are provided by the manufacturer. Automatic dispensers made available by some manufacturers permit more accurate control of the quantities of air-entraining agents used in the mix.

Properties of Air-Entrained Concrete.

Workability. Entrained air improves the workability of concrete. It is particularly effective in lean mixes and in mixes with angular and poorly graded aggregates. This improved workability allows a significant reduction in water and sand content. The disconnected air voids also reduce segregation and bleeding of plastic concrete.

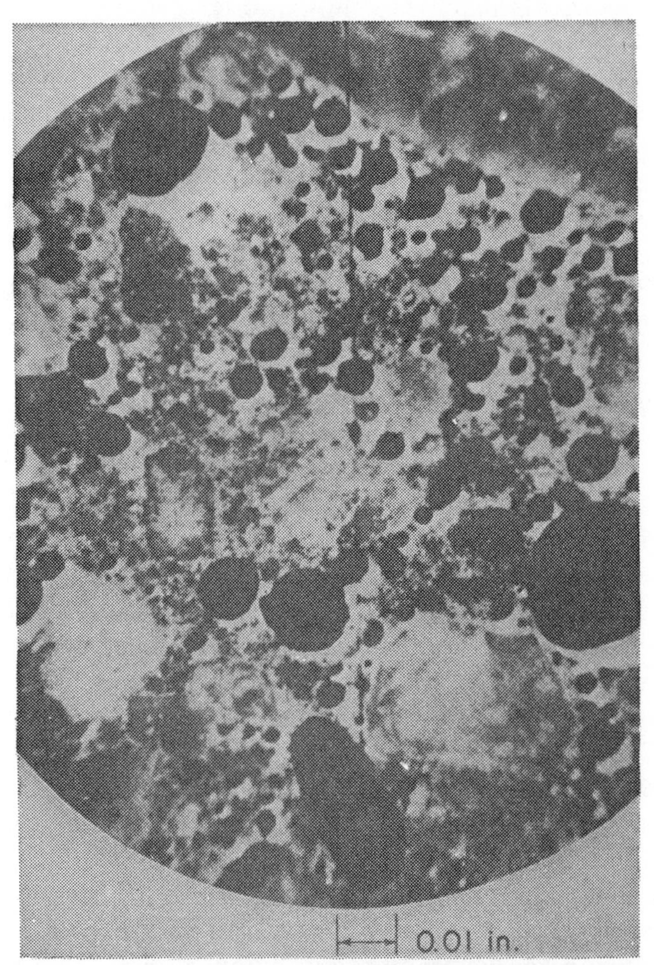

Figure 2–7. Polished section of air-entrained concrete magnified many times.

Freeze-thaw resistance. The freeze-thaw resistance of hardened concrete is significantly improved by the use of intentionally entrained air. As the water in concrete freezes, it expands, causing pressure that can rupture concrete. The entrained air voids act as reservoirs for excess water forced into them, thus relieving pressure and preventing damage to the concrete.

Resistance to de-icers. Entrained air is effective for preventing scaling caused by de-icing chemicals used for snow and ice removal. The use of air-entrained concrete is recommended for all concretes that come in any contact with de-icing chemicals.

Sulfate resistance. The use of entrained air improves the sulfate resistance of concrete as shown in figure 2–8. Concrete made with a low water-cement ratio, entrained air, and cement having a low tricalcium aluminate content will be most resistant to attack from sulfate soil waters or seawater.

Strength. Strength of air-entrained concrete depends principally upon the voids-cement ratio. For this ratio, "voids" is defined as the total volume of water plus air (entrained and entrapped). For a constant air content, strength varies inversely with the water-cement ratio. As air content is increased, a given strength generally may be maintained by holding to a constant voids-cement ratio by reducing amount of mixing water, increasing amount of cement, or both. Some reduction in strength may accompany air entrainment, but this is often minimized since air-entrained concretes have lower water-cement ratios than non-air-entrained concretes having the same slump. In some cases, however, it may be difficult to attain high strength with air-entrained concrete. Examples are when slumps are maintained constant as concrete temperatures rise and when certain aggregates are used.

Abrasion resistance. Abrasion resistance of air-entrained concrete is about the same as that of non-air-entrained concrete of the same compressive strength. Abrasion resistance increases as the compressive strength increases.

Watertightness. Air-entrained concrete is more watertight than non-air-entrained concrete,

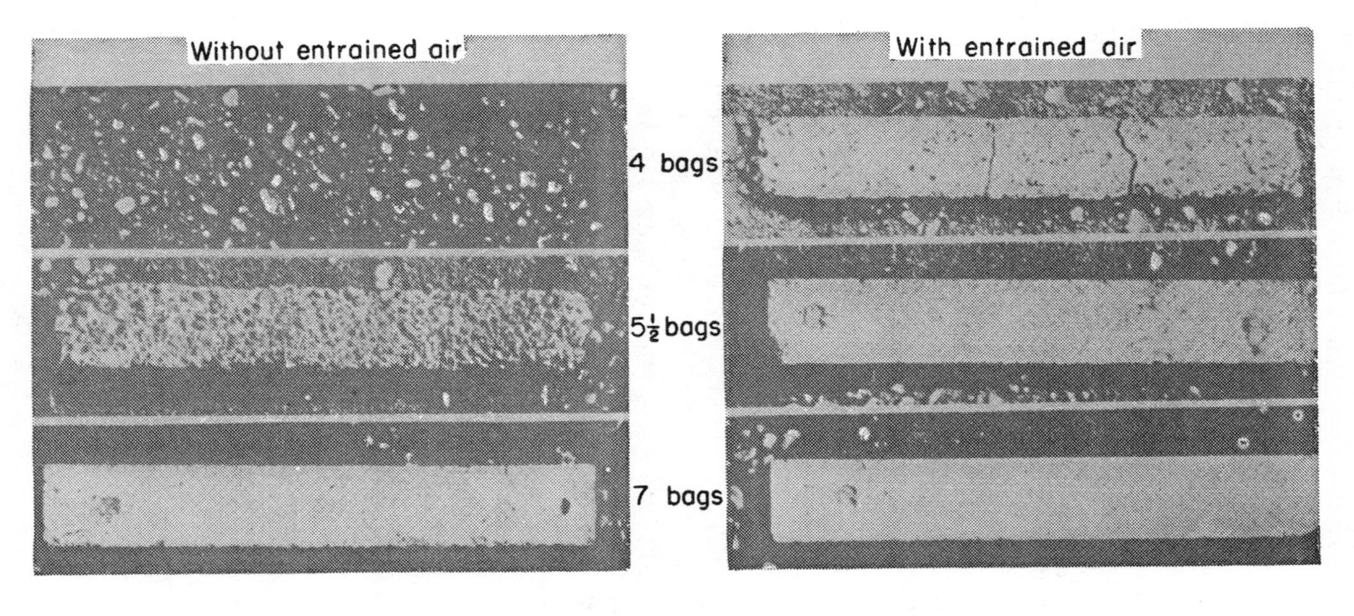

Figure 2–8. ***Effect of entrained air on performance of concrete specimens after 5 years of exposure to a sulfate soil.***

since entrained air inhibits the formation of inter-connected capillary channels. Air-entrained concrete should be used where watertightness is desired.

Air-Entraining Materials. The entrainment of air in concrete may be accomplished by using air-entraining cement, by adding an air-entraining agent at the mixer or by a combination of both methods. Air-entraining cements should meet the specifications in ASTM C175. Commercial air-entraining admixtures, which are manufactured from a variety of materials, may be added at the mixer and should comply with the specifications in ASTM C260. Adequate control is required to ensure the proper air content at all times.

Factors Affecting Air Content.

Aggregate gradation and cement content. Aggregate gradation and the cement content of a mix have a significant effect on the air content of both air-entrained and non-air-entrained concrete. For aggregate sizes smaller than 1½ inches, the air content increases sharply as the aggregate size decreases due to the increase in mortar volume. As the cement content increases, the air content decreases with the normal range of cement content.

Fine aggregate content. The percentage of entrained air in concrete is affected by the fine aggregate content of the mix. Increasing the amount of fine aggregate causes more air to be entrained for a given amount of air-entraining cement or admixture.

Slump and vibration. Slump and vibration affect the air content of air-entrained concrete. The greater the slump, the larger the percent reduction in air content during vibration. At all slumps, however, even 15 seconds of vibration causes a considerable reduction in air content.

Even so, if vibration is properly applied, the air lost will consist mostly of large bubbles and little of the intentionally entrained air will be lost.

Concrete temperature. Less air is entrained as the temperature of the concrete increases. The effect of temperature becomes more pronounced as the slump is increased.

Mixing action. Mixing action is the most important factor in the production of entrained air in concrete. The amount of entrained air varies with the type and condition of the mixer, the amount of concrete being mixed, and the rate of mixing. Stationary mixers and transit mixers may produce concretes with significant differences in the amounts of air entrained. Mixers loaded to less than capacity may produce increases in air content, and decreases may result from overloading. Generally, more air is entrained as the speed of mixing is increased.

Admixtures and coloring agents. Certain admixtures and coloring agents may reduce the amount of entrained air. This is particularly true of fly ash with high percentages of carbon. If calcium chloride is used, it should be added separately to the mix in solution form to prevent a chemical reaction with some air-entraining admixtures.

Premature finishing. Permature finishing operations may cause excess water from bleeding to be worked into the top surface of concrete. The surface zones may become low in entrained air and thus susceptible to scaling.

Recommended Air Contents. The amount of air to be used in air-entrained concrete should be flexible and suited to the particular need. It should depend on the type of structure, climatic conditions, number of freeze-thaw-cycles, extent of exposure to de-icers and aggressive soils or waters, and to some extent the strength of concrete. Field practice has shown that the amount of air indicated in table 2–4 should be specified for air-entrained concrete to provide adequate resistance to deterioration from freeze-thaw cycles and de-icing chemicals. Air contents are expressed in terms of percent by volume of the concrete, although air is entrained only in the mortar. In the field, a relatively constant amount of about 9 percent of air in the mortar fraction of the concrete (which corresponds to roughly 4½ percent by volume of the concrete) should provide the recommended air content for durability regardless of changes in cement content, maximum size of aggregate, consistency, and type of coarse aggregate. For exposure to extremely severe conditions, it is desirable to design air-entrained concrete for the highest air contents recommended in table 2–4. For certain concretes, such as those having high cement contents, low water contents, and consistencies below about one-inch slump, the level of air content recommended in table 2–4 is high. When entrained air is not required for protection against freeze-thaw or de-icers, the air content given in table 2–4 may be reduced by about one-third.

Tests for Air Content. Available methods for determining air entrainment in freshly mixed con-

*Table 2–4. Recommended Air Contents for Concretes Subject to Severe Exposure Conditions**

Maximum-size coarse aggregate, in.	Air content, percent by volume**
1½, 2, or 2½	5 ±1
¾ or 1	6 ±1
⅜ or ½	7½±1

*For structural lightweight concrete, add 2 percent to the values to allow for entrapped air in the aggregate particles, a range of ±1½ percent is permissible.

**The air content of the mortar fraction of the concrete should be about 9 percent.

crete measure only air volume and not the air void characteristics. This has been shown to be generally indicative of the adequacy of the air void system when using air-entraining materials meeting ASTM specifications. Tests should be made regularly during construction with samples taken immediately after discharge from the mixer and also from the concrete after it has been placed and consolidated. The following methods for determining the air content of freshly mixed concrete have been standardized:

Pressure method. This test (ASTM C231) is practical for field testing of all concretes except those made with highly porous and lightweight aggregates.

Volumetric method. This test (ASTM C173) is practical for field testing of all concretes, but is particularly useful for concretes made with lightweight and porous aggregates.

Gravimetric method. This test (ASTM C138) is impractical as a field test method since it requires accurate knowledge of specific gravities and absolute volumes of concrete ingredients but can be satisfactorily used in the laboratory.

Other Admixtures

Water-Reducing Admixtures. A water-reducing admixture is a material used for the purpose of reducing the quantity of mixing water required to produce concrete of a given consistency. These admixtures increase the slump for a given water content.

Retarding Admixtures. Retarders are sometimes used in concrete to reduce the rate of hydration to permit the placement and consolidation of concrete before the initial set. These admixtures are also used to offset the accelerating effect of hot weather on the setting of concrete. These admixtures generally fall in the categories of fatty acids, sugars, and starches.

Accelerating Admixtures. Accelerating admixtures accelerate the setting and the strength development of concrete. Calcium chloride is the most commonly used accelerator. It should be added in solution form as part of the mixing water and should not exceed 2 percent by weight of cement. Calcium chloride or other admixtures containing soluble chlorides should not be used in prestressed concrete, concrete containing embedded aluminum, concrete in permanent contact with galvanized steel, or concrete subjected to alkali-aggregate reaction or exposed to soils or water containing sulfates.

Pozzolans. Pozzolans are siliceous or siliceous and aluminous materials which combine with calcium hydroxide to form compounds possessing cementitious properties. The properties of pozzolans and their effects on concrete vary considerably. Before one is used it should be tested in order to determine its suitability.

Workability Agents. It is often necessary to improve the workability of fresh concrete. Workability agents frequently used include entrained air, certain organic materials, and finely divided materials. Fly ash and natural pozzalons used should conform to ASTM C618.

Dampproofing and Permeability-Reducing Agents. Dampproofing admixtures, usually water-repellant materials, are sometimes used to reduce the capillary flow of moisture through concrete that is in contact with water or damp earth. Permeability-reducing agents are usually either water-repellents or pozzolans.

Grouting Agents. The properties of portland cement grouts are altered by the use of various air-entraining admixtures, accelerators, retarders, workability agents, and so on, in order to meet the needs of a specific application.

Gas-Forming Agents. Gas-forming materials may be added to concrete or grout in very small quantities to cause a slight expansion prior to hardening in certain applications. However, while hardening, the concrete or grout made with gas-forming material has a decrease in volume equal to or greater than that for normal concrete or grout.

3

Proportioning Concrete Mixtures

BOOK METHOD

In arriving at the proportional quantities of cement, water and aggregate for a concrete mix, one of three methods (book, trial batch, or absolute volume) is commonly used. The book method is a theoretical procedure in which established data is used to determine mix proportions. Due to the variation of the materials (aggregates) used, mixes arrived at by the book method require adjustment in the field following the mixing of trial batches and testing. Concrete mixtures should be designed to give the most economical and practical combination of the materials that will produce the necessary workability in the fresh concrete and the required qualities in the hardened concrete.

Selecting Mix Characteristics

Certain information must be known before a concrete mixture can be proportioned. The size and shape of structural members, the concrete strength required, and the exposure conditions must be determined. The water-cement ratio, aggregate characteristics, amount of entrained air, and slump are significant factors in the selection of the appropriate concrete mixture.

Water-Cement Ratio. In arriving at the water-cement ratio, the requirements of strength, durability, and watertightness of the hardened concrete must be considered. These factors are usually specified by the engineer in the design of the structure or assumed for purposes of arriving at tentative mix proportions. It is important to remember that a change in the water-cement ratio changes the characteristics of the hardened concrete. Selection of a suitable water-cement ratio is made from table 3–1 for various exposure conditions. Note that the quantities are the recommended maximum permissible water-cement ratios. As indicated in table 3–1, under certain conditions the water-cement ratio should be selected on the basis of concrete strength. In such cases, if possible, tests should be made with job materials to determine the relationship between water-cement ratio and strength. If laboratory test data or experience records for this relationship cannot be obtained, the necessary water-cement ratio may be estimated from figures 3–1 and 3–2, the lower edge of the applicable strength band curve should be used, and the desired design strength of the concrete should be increased by 15 percent according to ACI requirements. If flexural strength rather than compressive strength is the basis for design as in pavements, tests should be made to determine the relationship between water-cement ratio and flexural strength. An approximate relationship between flexural and compressive strength is—

$$f'_c = \left(\frac{R}{K} \right)^2$$

where f'_c = compressive strength, in psi

R = flexural strength (modulus of rupture), in psi, third-point loading

Table 3-1. ACI Recommended Maximum Permissible Water-Cement Ratios

Type of structures	In air
A. Thin sections such as reinforced piles and pipe	5.5
B. Bridge decks	5
C. Thin sections such as railings, curbs, sills, ledges, ornamental or architectural concrete, and all sections with less than 1-in. concrete cover over reinforcement	5.5
D. Moderate sections, such as retaining walls, abutments, piers, girders, beams	6
E. Exterior portions of heavy (mass) sections	6.5
F. Concrete deposited by tremie under water	
G. Concrete slabs laid on the ground	6
H. Pavements	5.5
I. Concrete protected from the weather, interiors of buildings, concrete below ground	††
J. Concrete which will later be protected by enclosure or backfill but which may be exposed to freezing and thawing for several years before such protection is offered	6

*Adapted from Recommended Practice for Selecting Proportions for Concrete (ACI 613–54).

**Air-entrained concrete should be used under all conditions involving severe exposure and may be used under mild exposure conditions to improve workability of the mixture.

†Soil or groundwater containing sulfate concentrations of more than 0.2 per cent. For moderate sulfate resistance. the tricalcium aluminate content of the cement should be

K = a constant, usually between 8 and 10.

*for Different Types of Structures and Degrees of Exposure**

Exposure conditions**				
Severe wide range in temperature or frequent alternations of freezing and thawing (air-entrained concrete only) (gallons/sack)		Mild temperature rarely below freezing, or rainy, or arid (gallons/sack)		
At water line or within range of fluctuating water level or spray		In air	At water line or within range of fluctuating water level or spray	
In fresh water	In sea water or in contact with sulfates†		In fresh water	In sea water or in contact with sulfates†
5	4.5	6	5.5	4.5
5	4.5	5.5	5.5	5
--------	--------	6	5.5	--------
[illegible]	5	††	6	5
[illegible].5	5	††	6	5
[illegible]	5	--------	5	5
--------	--------	††	--------	--------
--------	--------	6	--------	--------
--------	--------	††	--------	--------
--------	--------	††	--------	--------

limited to 8 per cent. and for high-sulfate resistance to 5 per cent. At equal cement contents. air-entrained concrete is significantly more resistant to sulfate attack than non-air-entrained concrete.

††Water-cement ratio should be selected on basis of strength and workability requirements. but minimum cement content should not be less than 470 lb. per cubic yard.

In cases where both exposure conditions and strength must be considered, the lower of the two indicated water-cement ratios should be used.

Aggregate.

Fine aggregate. Fine aggregate fills the spaces in the coarse aggregate and increases the workability of the mix. In general, aggregates which do not have a large deficiency or an excess of any size and give a smooth grading curve produce the most satisfactory mix. Fine aggregate grading and fineness modulus are discussed in paragraph 2-5*e*(1) and (2).

Coarse aggregate. The largest size aggregate which is practical should be used. The larger the maximum size of the coarse aggregate, the less mortar and paste will be necessary. It follows that the larger the coarse aggregate, the less water and cement will be required for a given quality of concrete. The maximum size aggregate should not exceed one-fifth the minimum dimension of the member, or three-fourths of the clear space between reinforcing bars. For payment or floor slabs, the maximum size aggregate should not exceed one-third the slab thickness. The maximum size of coarse aggregate that produces concrete of maximum strength for a given cement content depends upon aggregate source as well as aggregate shape and grading. For many aggregates, this "optimum" maximum size is about ¾ inch.

Entrained Air. Entrained air should be used in all concrete exposed to freezing and thawing and may be used for mild exposure conditions to improve workability. It is recommended for all paving concrete regardless of climatic conditions. The recommended total air contents for air-entrained concretes are shown in table 3-2. When mixing water is held constant, the entrainment of

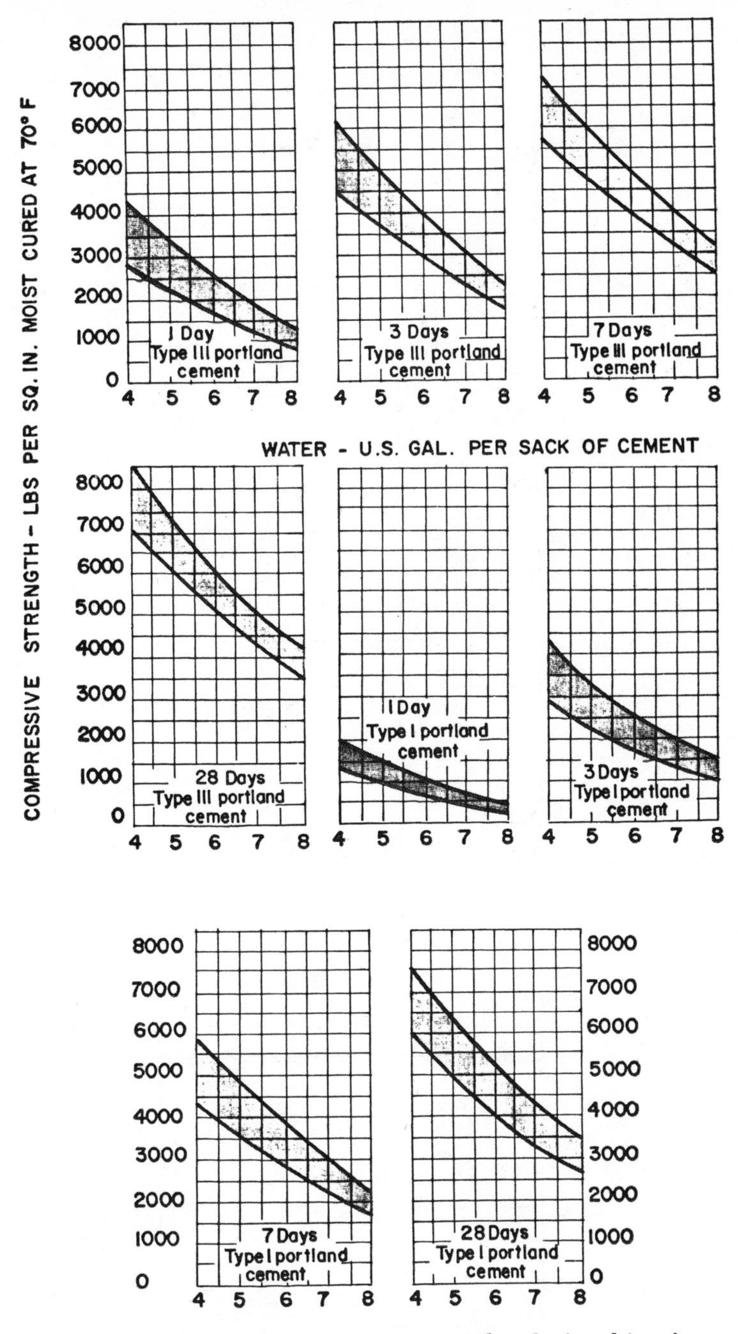

Figure 3–1. Age-compressive strength relationships for types I and III non-air-entrained portland cement.

*Table 3-2. Approximate Mixing Water Requirements for Different Slumps and Maximum Sizes of Aggregates**

Maximum size of aggregate, in.	Air-entrained concrete				Non-air-entrained concrete			
	Recommended average total air content, per cent†	Slump, in.			Approximate amount of entrapped air, per cent	Slump, in.		
		1 to 2	3 to 4	5 to 6		1 to 2	3 to 4	5 to 6
		Water, gal. per cu.yd. of concrete**				Water, gal. per cu.yd. of concrete**		
⅜	7.5	37	41	43	3.0	42	46	49
½	7.5	36	39	41	2.5	40	44	46
¾	6.0	33	36	38	2.0	37	41	43
1	6.0	31	34	36	1.5	36	39	41
1½	5.0	29	32	34	1.0	33	36	38
2	5.0	27	30	32	0.5	31	34	36
3	4.0	25	28	30	0.3	29	32	34
6	3.0	22	24	26	0.2	25	28	30

*Adapted from Recommended Practice for Selecting Proportions for Concrete (ACI 613-54).

**These quantities of mixing water are for use in computing cement factors for trial batches. They are maximums for reasonably well-shaped angular coarse aggregates graded within limits of accepted specifications.

†Plus or minus 1 per cent.

air will increase slump. When cement content and slump are held constant, less mixing water is required; the resulting decrease in the water-cement ratio helps to offset possible strength decreases and results in improvements in other paste properties such as permeability. Hence, the strength of air-entrained concrete may equal, or nearly equal, that of non-air-entrained concrete when their cement contents and slumps are the same.

Slump. The slump test is generally used as a measure of the consistency of concrete. It should not be used to compare mixes with wholly different proportions or mixes with different kinds of sizes of aggregates. When used to test different batches of the same mixture, changes in slump indicate changes in materials, mix proportions, or

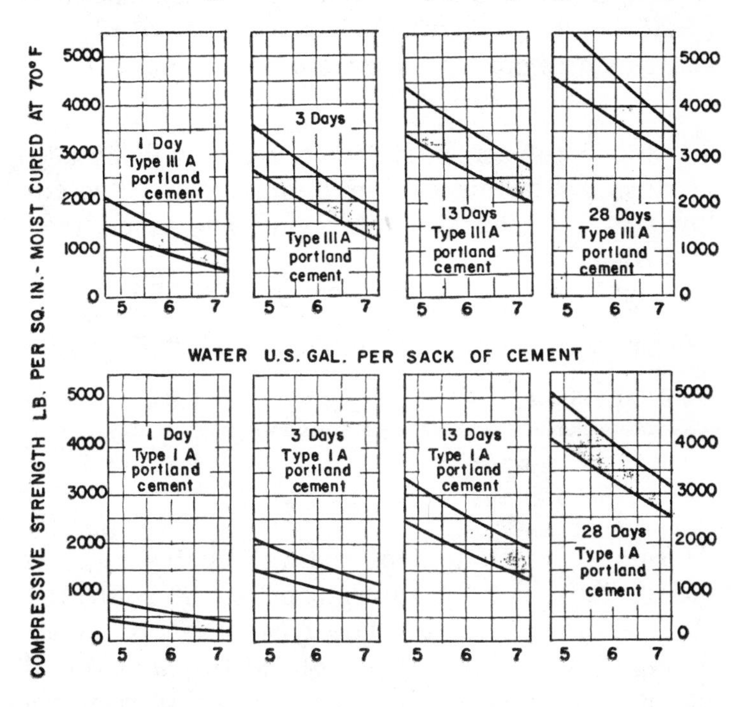

Figure 3–2. Age-compressive strength relationships for types I and III air-entrained portland cement.

water content. Acceptable slump ranges are indicated in table 3–3.

Mix Proportions. Knowing the water-cement ratio, slump, maximum size of aggregate, and fineness modulus, tables of trial mixes such as tables 3–4 and 3–5 can be used to determine the proportions of trial mixes. The quantities in tables 3–4 and 3–5 are based on concrete having a slump of 3 to 4 inches, with well graded aggregates having a specific gravity of 2.65. For other conditions it is necessary to adjust the quantities in accordance with the footnotes.

Example. The mix proportions are to be determined for a water-cement ratio of 6 gallons per sack, maximum aggregate size of 1 inch, aircontent of 6 ± 1 percent, and a slump of 3 to 4 inches. The fine aggregate has a fineness modulus of 2.50 and a moisture content of 5 percent. The coarse aggregate has a moisture content of 1 percent.

Table 3–5 for air-entrained concrete will be used because of the air content requirement. The following quantities per cubic yard are taken from table 3–5:

Cement	5.7 sacks =	535 pounds
Water	34 gallons =	285 pounds
Fine aggregate		1,040 pounds
Coarse aggregate		1,940 pounds
Total		3,800 pounds

The moisture content of the aggregates must be considered since the tables are for the saturated, surface-dry condition. The free moisture in the fine aggregate is 5 percent of 1,040 pounds or 52 pounds (use 50 pounds). The free moisture in the coarse aggregate is 1 percent of 1,940 pounds or approximately 19 pounds (use 20 pounds). The corrected weights per cubic yard of concrete are—

Cement		535 pounds
Water	285 — 50 — 20 =	215 pounds
Fine aggregate	1,040 + 50 =	1,090 pounds
Coarse aggregate	1,940 + 20 =	1,960 pounds
Total		3,800 pounds

When these quantities were mixed the consistency was such that the slump was approximately 1 inch. Additional water (25 pounds) was added to bring the slump to slightly more than 3 inches. The unit weight was measured to be 145 pounds per cubic foot. Since the unit weight of the concrete was 145 pounds per cubic foot and the total weight of concrete was 3,800 + 25 = 3,825 pounds, the volume of concrete was 26.4 cubic feet. This is slightly less than 1 cubic yard (27 cubic feet). The principal reason for this discrep-

*Table 3-3. Recommended Slumps for Various Types of Construction**

Type of construction	Slump, inches	
	Maximum	Minimum
Reinforced foundation walls and footings	6	3
Unreinforced footings, caissons, and substructure walls	4	1
Reinforced slabs, beams, and walls	6	3
Building columns	6	4
Pavements	3	1
Heavy mass construction	3	1
Bridge decks	4	3
Sidewalk, driveway, and slabs on ground	6	3

*When high-frequency vibrators are used, the values may be decreased approximately one-third, but in no case should the slump exceed 6 inches.

*Table 3–4. Suggested Trial Mixes for Non-Air-Entrained Concrete of Medium Consistency with 3- to 4-Inch Slump**

Water-cement ratio Gal per sack	Maximum size of aggregate inches	Air content (entrapped air) per cent	Water gal per cu yd of concrete
4.5	3/8	3	46
	1/2	2.5	44
	3/4	2	41
	1	1.5	39
	1 1/2	1	36
5.0	3/8	3	46
	1/2	2.5	44
	3/4	2	41
	1	1.5	39
	1 1/2	1	36
5.5	3/8	3	46
	1/2	2.5	44
	3/4	2	41
	1	1.5	39
	1 1/2	1	36
6.0	3/8	3	46
	1/2	2.5	44
	3/4	2	41
	1	1.5	39
	1 1/2	1	36
6.5	3/8	3	46
	1/2	2.5	44
	3/4	2	41
	1	1.5	39
	1 1/2	1	36
7.0	3/8	3	46
	1/2	2.5	44
	3/4	2	41
	1	1.5	39
	1 1/2	1	36
7.5	3/8	3	46
	1/2	2.5	44
	3/4	2	41
	1	1.5	39
	1 1/2	1	36
8.0	3/8	3	46
	1/2	2.5	44
	3/4	2	41
	1	1.5	39
	1 1/2	1	36

*See footnote at end of table.

Cement sacks per cu yd of concrete	With fine sand—fineness modulus = 2.50		
	Fine aggregate per cent of total aggregate	Fine aggregate lb per cu yd of concrete	Coarse aggregate lb per cu yd of concrete
10.3	50	1240	1260
9.8	42	1100	1520
9.1	35	960	1800
8.7	32	910	1940
8.0	29	880	2110
9.2	51	1330	1260
8.8	44	1180	1520
8.2	37	1040	1800
7.8	34	990	1940
7.2	31	960	2110
8.4	52	1390	1260
8.0	45	1240	1520
7.5	38	1090	1800
7.1	35	1040	1940
6.5	32	1000	2110
7.7	53	1440	1260
7.3	46	1290	1520
6.8	39	1130	1800
6.5	36	1080	1940
6.0	33	1040	2110
7.1	54	1480	1260
6.8	46	1320	1520
6.3	39	1190	1800
6.0	37	1120	1940
5.5	34	1070	2110
6.6	55	1520	1260
6.3	47	1360	1520
5.9	40	1200	1800
5.6	37	1150	1940
5.1	34	1100	2110
6.1	55	1560	1260
5.9	48	1400	1520
5.5	41	1240	1800
5.2	38	1190	1940
4.8	35	1130	2110
5.7	56	1600	1260
5.5	48	1440	1520
5.1	42	1280	1800
4.9	39	1220	1940
4.5	35	1160	2110

Table 3-4. Continued

With average sand—fineness modulus = 2.75		
Fine aggregate percent of total aggregate	Fine aggregate lb per cu yd of concrete	Coarse aggregate lb per cu yd of concrete
52	1310	1190
45	1170	1450
37	1030	1730
34	980	1870
32	960	2030
54	1400	1190
46	1250	1450
39	1110	1730
36	1060	1870
34	1040	2030
55	1460	1190
47	1310	1450
40	1160	1730
37	1110	1870
35	1080	2030
56	1510	1190
48	1360	1450
41	1200	1730
38	1150	1870
36	1120	2030
57	1550	1190
49	1390	1450
42	1240	1730
39	1190	1870
36	1150	2030
57	1590	1190
50	1430	1450
42	1270	1730
39	1220	1870
37	1180	2030
58	1630	1190
50	1470	1450
43	1310	1730
40	1260	1870
37	1210	2030
58	1670	1190
51	1520	1450
44	1350	1730
41	1290	1870
38	1250	2030

*Increase or decrease water per cubic yard by 3 per cent for each increase or
For manufactured fine aggregate, increase percentage of fine aggregate by 3 and water
decrease percentage of fine aggregate by 3 and water by 8 lb. per cubic yard of concrete.

With coarse sand—fineness modulus = 2.90		
Fine aggregate percent of total aggregate	Fine aggregate lb per cu yd of concrete	Coarse aggregate lb per cu yd of concrete
54	1350	1150
47	1220	1400
39	1080	1680
36	1020	1830
33	1000	1990
56	1440	1150
48	1300	1400
41	1160	1680
38	1100	1830
35	1080	1990
57	1500	1150
49	1360	1400
42	1210	1680
39	1150	1830
36	1120	1990
57	1550	1150
50	1410	1400
43	1250	1600
39	1190	1830
37	1160	1990
58	1590	1150
51	1440	1400
43	1290	1680
40	1230	1830
37	1190	1990
59	1630	1150
51	1480	1400
44	1320	1680
41	1260	1830
38	1220	1990
59	1670	1150
52	1520	1400
45	1370	1600
42	1300	1830
39	1250	1990
60	1710	1150
53	1560	1400
45	1400	1680
42	1330	1830
39	1280	1990

decrease of 1 in. in slump, then calculate quantities by absolute volume method.
by 17 lb. per cubic yard of concrete. For less workable concrete, as in pavements,

Table 3–5. Suggested Trial Mixes for Air-Entrained

Water-cement ratio Gal per sack	Maximum size of aggregate inches	Air Content (entrapped air) per cent	Water gal per cu yd of concrete
4.5	3/8	7.5	41
	1/2	7.5	39
	3/4	6	36
	1	6	34
	1 1/2	5	32
5.0	3/8	7.5	41
	1/2	7.5	39
	3/4	6	36
	1	6	34
	1 1/2	5	32
5.5	3/8	7.5	41
	1/2	7.5	39
	3/4	6	36
	1	6	34
	1 1/2	5	32
6.0	3/8	7.5	41
	1/2	7.5	39
	3/4	6	36
	1	6	34
	1 1/2	5	32
6.5	3/8	7.5	41
	1/2	7.5	39
	3/4	6	36
	1	6	34
	1 1/2	5	32
7.0	3/8	7.5	41
	1/2	7.5	39
	3/4	6	36
	1	6	34
	1 1/2	5	32
7.5	3/8	7.5	41
	1/2	7.5	39
	3/4	6	36
	1	6	34
	1 1/2	5	32
8.0	3/8	7.5	41
	1/2	7.5	39
	3/4	6	36
	1	6	34
	1 1/2	5	32

*See footnote at end of table.

*Concrete of Medium Consistency with 3- to 4-Inch Slump**

Cement sacks per cu yd of concrete	With fine sand—fineness modulus = 2.50		
	Fine aggregate per cent of total aggregate	Fine aggregate lb per cu yd of concrete	Coarse aggregate lb per cu yd of concrete
9.1	50	1250	1260
8.7	41	1060	1520
8.0	35	970	1800
7.8	32	900	1940
7.1	29	870	2110
8.2	51	1330	1260
7.8	43	1140	1520
7.2	37	1040	1800
6.8	33	970	1940
6.4	31	930	2110
7.5	52	1390	1260
7.1	44	1190	1520
6.5	38	1090	1800
6.2	34	1010	1940
5.8	32	970	2110
6.8	53	1430	1260
6.5	45	1230	1520
6.0	38	1120	1800
5.7	35	1040	1940
5.3	32	1010	2110
6.3	54	1460	1260
6.0	45	1260	1520
5.5	39	1150	1800
5.2	36	1080	1940
4.9	33	1040	2110
5.9	54	1500	1260
5 6	46	1300	1520
5.1	40	1180	1800
4.9	36	1100	1940
4.6	33	1060	2110
5.5	55	1530	1260
5.2	47	1330	1520
4.8	40	1210	1800
4.5	37	1140	1940
4.3	34	1090	2110
5.1	55	1560	1260
4.9	47	1360	1520
4.5	41	1240	1800
4.3	37	1160	1940
4.0	34	1110	2110

Table 3-5. ***Continued***

With average sand—fineness modulus = 2.75		
Fine aggregate percent of total aggregate	Fine aggregate lb per cu yd of concrete	Coarse aggregate lb per cu yd of concrete
53	1320	1190
44	1130	1450
38	1040	1730
34	970	1870
32	950	2030
54	1400	1190
46	1210	1450
39	1110	1730
36	1040	1870
33	1010	2030
55	1460	1190
46	1260	1450
40	1160	1730
37	1080	1870
34	1050	2030
56	1500	1190
47	1300	1450
41	1190	1730
37	1110	1870
35	1090	2030
56	1530	1190
48	1330	1450
41	1220	1730
38	1150	1870
36	1120	2030
57	1570	1190
49	1370	1450
42	1250	1730
38	1170	1870
36	1140	2030
57	1600	1190
49	1400	1450
43	1280	1730
39	1210	1870
37	1170	2030
58	1630	1190
50	1430	1450
43	1310	1730
40	1230	1870
37	1190	2030

•Increase or decrease water per cubic yard by 3 per cent for each increase or
For manufactured fine aggregate, increase percentage of fine aggregate by 3 and water
decrease percentage of fine aggregate by 3 and water by 8 lb. per cubic yard of concrete.

With coarse sand—fineness modulus = 2.90		
Fine aggregate percent of total aggregate	Fine aggregate lb per cu yd of concrete	Coarse aggregate lb per cu yd of concrete
54	1360	1150
46	1180	1400
39	1090	1680
36	1010	1830
33	990	1990
56	1440	1150
47	1260	14000
41	1160	1630
37	1080	1830
35	1050	1990
57	1500	1150
48	1310	1400
42	1210	1680
38	1120	1830
35	1090	1990
57	1540	1150
49	1350	1400
42	1240	1680
39	1150	1830
36	1130	1990
58	1570	1150
50	1380	1400
43	1270	1680
39	1190	1830
37	1160	1990
58	1610	1150
50	1420	1400
44	1300	1680
40	1210	1830
37	1180	1990
59	1640	1150
51	1450	1400
44	1330	1680
41	1250	1830
38	1210	1990
59	1670	1150
51	1480	1400
44	1360	1680
41	1270	1830
38	1230	1990

decrease of 1 in. in slump, then calculate quantities by absolute volume method
by 17 lb. per cubic yard of concrete. For less workable concrete, as in pavements,

ancy is that the specific gravity values of the aggregates were probably different from those assumed in tables 3–4 and 3–5. In small adjustments it may be assumed that the unit weight of the concrete remains essentially constant and that the amount of water required per cubic yard of concrete remains constant. The adjusted water requirement is—

$$\frac{27}{26.4} \times 310 \text{ pounds} = 315 \text{ pounds} = 38 \text{ gallons}$$

Note that the 310 pounds used is the total amount of water needed, 285 + 25 pounds. The adjusted cement requirement is

$$\frac{38 \text{ gallons}}{\frac{6 \text{ gallons}}{\text{sack}}} = 6.3 \text{ sacks} = 595 \text{ pounds}$$

The weight of materials per cubic yard of concrete must total 145 × 27 = 3,910 pounds, approximately. The total weight of aggregates must therefore be 3,910 —315 —595 = 3,000 pounds. Table 3–5 indicates that 35 percent of this should be fine aggregate.

The adjusted trial mix proportions (per cubic yard) are therefore:

Cement	6.3	sacks
Water	38	gallons
Fine aggregate	1,050	pounds
Coarse aggregate	1,950	pounds

TRIAL BATCH METHOD

The trial batch method of mix design utilizes the actual materials in arriving at mix proportions instead of the tables of trial mixes (tables 3–4 and 3–5). When the quality of the concrete mixture is specified in terms of the water-cement ratio, the trial batch procedure consists essentially of combining a paste (water, cement, and, generally, entrained air) of the correct proportions with the necessary amounts of fine and coarse aggregates to produce the required slump and workability. Quantities per sack and/or per cubic yard are then calculated. It is important to use representative samples of the aggregates, cement, water, and air-entraining admixture, if used. The aggregates should be pre-wetted; allowed to dry to a saturated, surface-dry condition; and placed in covered containers to keep them in this condition until use. This procedure simplifies calculations and eliminates error caused by variations in aggregate moisture content. The size of the trial batch is dependent on the equipment and the test specimens to be made. Batches using 10 to 20 pounds of cement may be adequate, although larger batches will produce more accurate data. Machine mixing is recommended since it more nearly represents job conditions; it is mandatory if the concrete is to contain entrained air.

Example Using Trial Batch Procedure

The mix proportions are to be determined for concrete which will be used in a retaining wall that will be exposed to fresh water in a severe climate. A compressive strength of 3,000 psi at 28 days is required. The minimum thickness of the wall is 8 inches and 2 inches of concrete must cover the reinforcement. All trial mix data will be entered in the appropriate blanks on the trial mix data worksheet, figure 3–3.

Line D of table 3–1 indicates that a maximum water-cement ratio of 5.5 gallons per sack will satisfy the exposure requirements. Using type IA (air-entrained) portland cement and a compressive strength of 3,450 psi (3,000 psi +15 percent), figure 3–2 indicates that a maximum water-cement ratio of approximately 5.75 gallons per sack will satisfy the strength requirements. In order to meet both specifications a water-cement ratio of 5.5 gallons per sack is selected. Since the maximum size of coarse aggregate must not exceed one-fifth the minimum thickness of the wall, nor three-fourths of the clear space between reinforcement and the surfaces, the maximum size of coarse aggregate is chosen as 1½ inches. Because of the severe exposure conditions, the concrete should contain entrained air. From table 3–2, the recommended air content is 5 ± 1 percent. If we assume that the concrete will be consolidated by vibration, table 3–3 indicates a recommended slump of from 2 to 4 inches. The trial batch proportions are now determined. A batch containing 20 pounds of cement is chosen for convenience. The mixing water required is therefore

$$\frac{20}{94} \times 5.5 \frac{\text{gal}}{\text{sack}} \times 8.33 \frac{\text{lb}}{\text{gal}} = 9.8 \text{ pounds}$$

Representative samples of fine and coarse aggregates are selected and weighed. This is recorded in column (2) of figure 3–3. All of the measured quantities of cement, water, and air-entraining admixture are used. Fine and coarse aggregates are then added until a workable mixture having the proper slump is produced. Figure 3–4 indicates the appearance of fresh concrete with correct and incorrect amounts of mortar. The weight of material actually used is recorded in column (4). The weights for a 1-bag batch and per cubic yard are calculated and recorded in columns (5) and (6), respectively. The cement factor in bags per cubic yard is calculated and recorded as indicated in figure 3–3. The percentage of fine aggregate by weight of total aggregate is also included as is the yield of concrete in cubic feet per bag. The slump, air content, workability, and unit weight of concrete are determined and noted as shown. To determine the most economical proportions, additional trial batches should be made varying the percentage of fine aggregate. In each batch the water-cement ratio, aggregate gradations, air content, and slump are maintained approximately the same. Results of four such trial batches are summarized in table 3–6. For these mixes, the percentage of fine aggregate is plotted against the cement factor in figure 3–5. The minimum cement factor (5.72 sacks per cubic yard, use 5.7) occurs at a fine aggregate content of about 32 percent of total aggregate. Since the water-cement ratio is 5.5 gallons per sack and the unit weight of the concrete for an air content of 5 percent is about 144 pounds per cubic foot, the final quantities for the mix proportion (per cubic yard) are—

Cement		5.7 sacks =	535 pounds
Water 5.7 sacks × 5.5 gal/sack		31.5 gallons =	260 pounds
		Total	795 pounds
Concrete per cubic yard	144 x 27	=	3,890 pounds
Aggregates	3,890 — 795	=	3,095 pounds
Fine aggregate	.32 × 3,095	=	990 pounds
Coarse aggregate	3,095 — 990	=	2,105 pounds

*Table 3–6. Examples of Results of Laboratory Trial Mixes**

Batch no.	Slump, in.	Air content, per cent	Unit wt., pcf	Cement factor, sacks per cu. yd.	Fine aggregate, per cent of total aggregate	Workability
1	3	5.4	144	5.74	33.5	Excellent
2	2¾	4.9	144	5.91	27.4	Harsh
3	2½	5.1	144	5.84	35.5	Excellent
4	3¼	4.7	145	5.74	30.5	Good

*The water-cement ratio selected was 5.5 gal/sack.

CONCRETE

TRIAL MIX DATA

1. PROJECT NO: ______
2. STRUCTURE: Retaining Wall
3. EXPOSURE CONDITION:
   Severe or Moderate ✓ Mild ____
   In air ____
   In Fresh Water ✓
   In Sea Water ____
4. TYPE OF STRUCTURE (A-I): D
   MAX. W/C FOR EXPOSURE: 5.5 gal/sack
   MAX. W/C FOR WATERTIGHTNESS 5.5 gal/sack
5. TYPE OF CEMENT 1A
6. FINENESS MODULUS OF SAND 2.75
7. SPECIFIC GRAVITY:
   Sand 2.60
   Gravel 2.65
8. MAXIMUM SIZE AGGREGATE: 1½"
9. AIR CONTENT 5 % ± 1
10. DESIRED SLUMP RANGE
    MAX. 4 in. MIN. 2 in.
11. STRENGTH REQUIREMENT: 3450 psi
    W/C FOR STRENGTH: 5.75 gal/bag
    USE W/C 5.5 gal/sack
    gal/sack

DATA FOR TRIAL BATCH
(Saturated, surface-dry aggregates)

(1) Material	(2) Initial Wt. (lb)	(3) Final Wt. (lb)	(4) Wt. Used (lb)	(5) Wt. for 1-bag Batch	(6) Wt. per cu.yd.	(7) REMARKS
Cement	20.0	0	20.0	94.0	540	Cement Factor = 5.74 bag/cu.yd.
Water	9.8	0	9.8	46.1	265	31.8 gal/cu.yd.
Fine Agg.	66.2	27.9	38.3	180.0	1,035	33.5 % of total aggregate
Coarse Agg.	89.8	13.8	76.0	357.0	2,050	
Air-Entraining Admixture	.30 oz		TOTAL (T) = 677.1			

Measured Slump: 3 in. Air Content 5.4 % Workability: GOOD

Wt. Container concrete (lb.)	42.6
Wt. Container (lb.)	6.6
Wt. Concrete = A (lb.)	36.0
Vol. Container = B (cu.ft.)	.25

Unit Wt. of concrete = w = $\frac{A}{B}$

$\frac{A}{B} = \frac{36.0}{.25} = 144$ lb/cu.ft.

Yield = $\frac{T}{w} = \frac{677.1}{144} = 4.7$ cu.ft. bag

CALCULATIONS

Column (4) = Column (2) minus Column (3)

Column (5) = Column (4) times (94 ÷ Wt. of Cement Used)

$$\text{Yield} = \frac{\text{Total Wt. of material for 1-bag batch (T)}}{\text{Unit wt. of concrete (w)}}$$

Cement factor = 27 ÷ yield

Column (6) = Column (5) times cement factor

$$\text{Gal./cu.yd.} = \frac{\text{(Cement factor)} \times \text{(wt. of water for 1-bag batch)}}{\text{8.33 (lb. of water per gal.)}}$$

$$= \text{(Cement factor)} \times \text{(W/C)}$$

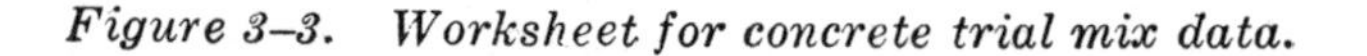

Figure 3–3. Worksheet for concrete trial mix data.

(a) A concrete mixture in which there is not sufficient cement-sand mortar to fill all the spaces between coarse aggregate particles. Such a mixture will be difficult to handle and place and will result in rough, honeycombed surfaces and porous concrete.

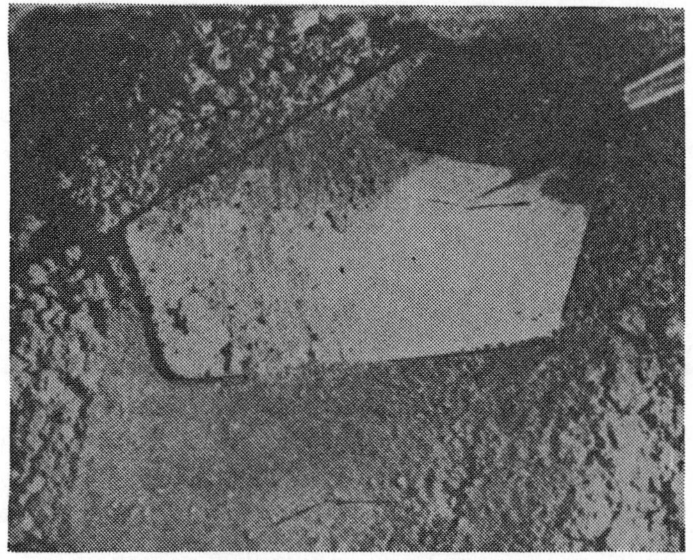

(b) A concrete mixture which contains correct amount of cement-sand mortar. With light troweling all spaces between coarse aggregate particles are filled with mortar. Note appearance on edges of pile. This is a good workable mixture and will give maximum yield of concrete with a given amount of cement.

Figure 3–4. Appearance of concrete mixes with correct and incorrect amounts of mortar.

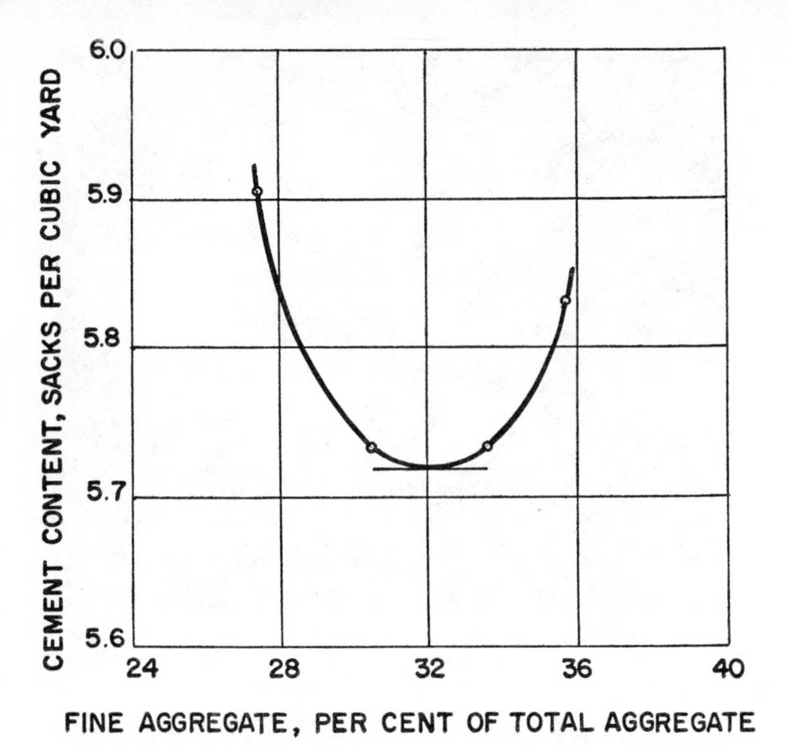

Figure 3–5. Relationship between percentage of fine aggregate and cement content for a given water-cement ratio and slump.

ABSOLUTE VOLUME METHOD

Concrete mixtures may be proportioned by using absolute volumes. This method is detailed in the American Concrete Institute (ACI) report, Recommended Practice for Selecting Proportions for Concrete (ACI 613–54). In this method, the water-cement ratio, slump air, content, and maximum size of aggregate are selected as before. In addition, the water requirement is estimated from table 3–2. Certain additional items must be known before calculations can be made. These are: the specific gravities of fine and coarse aggregates, the dry-rodded unit weight of coarse aggregate, and

Table 3–7. Volume of Coarse Aggregate Per Cubic Yard of Concrete

Maximum size of aggregate, in.	Fineness modulus of fine aggregate			
	2.40	2.60	2.80	3.00
	Coarse aggregate, cu. ft. per cu. yd.*			
⅜	13.5	13.0	12.4	11.9
½	15.9	15.4	14.8	14.3
¾	17.8	17.3	16.7	16.2
1	19.2	18.6	18.1	17.6
1½	20.2	19.7	19.2	18.6
2	21.1	20.5	20.0	19.4
3	22.1	21.6	21.1	20.5

*Volumes are based on aggregates in dry-rodded condition as described in Method of Test for Unit Weight of Aggregate (ASTM C29). These volumes are selected from empirical relationships to produce concrete with a degree of workability suitable for usual reinforced construction. For less workable concrete such as required for concrete pavement construction, they may be increased about 10 percent. When placement is to be by pump, they should be decreased about 10 percent.

the fineness modulus of the fine aggregate. If the maximum size of aggregate and the fineness modulus of the fine aggregate are known, the volume of dry-rodded coarse aggregate per cubic yard can be estimated from table 3–7. Then the quantities per cubic yard of water, cement, coarse aggregate, and air can be calculated. The sum of the absolute volumes of these materials in cubic feet is then subtracted from 27 to give the specific volume of fine aggregate.

Example Using Absolute Volume Method

The mix proportions are to be determined for the following conditions:

Maximum water-cement ratio	5.5 gallons per sack
Maximum size of aggregate	¾ inch
Air content	6 ± 1 percent
Slump	2 to 3 inches
Finenes modulus of fine aggregate	2.75
Specific gravity of portland cement	3.15
Specific gravity of fine aggregate	2.66

Specific gravity of coarse aggregate	2.61
Dry-rodded unit weight of coarse aggregate	104 pounds per cubic foot

The water requirement is estimated at about 34 gallons per cubic yard from table 3–2. Since the maximum water-cement ratio is 5.5 gallons per sack, the cement factor must be at least 6.2 sacks per cubic yard. The volume of dry-rodded coarse aggregate is estimated to be 16.9 cubic feet per cubic yard from table 3–7. Thus the weight of coarse aggregate is 16.9 x 104 = 1,758 pounds. The absolute volumes of these quantities of materials is now calculated by use of the relationship:

$$\text{absolute volume} = \frac{\text{weight of material}}{\text{specific gravity} \times \text{unit weight of water}}$$

The unit weight of water is 62.4 pounds per cubic foot, and water weighs approximately 8.33 pounds per gallon. Thus the absolute volumes are—

Cement

$$\frac{6.2 \times 94}{3.15 \times 62.4} = 2.97 \text{ cubic feet}$$

Water

$$\frac{34 \times 8.33}{1 \times 62.4} = 4.54 \text{ cubic feet}$$

Coarse aggregate

$$\frac{1{,}758}{2.61 \times 62.4} = 10.80 \text{ cubic feet}$$

Air

$$.06 \times 27 = 1.62 \text{ cubic feet}$$

$$\text{Total} = 19.93 \text{ cubic feet}$$

The absolute volume of the fine aggregate is—

$$27.00 - 19.93 = 7.07 \text{ cubic feet}$$

and its weight is—

$$7.07 \times 2.66 \times 62.4 = 1{,}170 \text{ pounds.}$$

These estimated quantities should be used for the first trial batch. Subsequent batches should be adjusted by maintaining the water-cement ratio constant and achieving the desired slump and air content.

Variation in Mixes

The proportions arrived at in determining mixes will vary somewhat depending upon which method is used. This occurs because of the empirical nature of these methods. It does not necessarily imply that one method is better than another. Each method begins by assuming certain needs or requirements and then proceeding to determine the other variables. Since the methods begin differently and use different procedures, the final proportions vary slightly. This is to be expected, and it further points out the necessity of trial mixes in determining the final mix proportions.

4

Form Design And Construction

PRINCIPLES

Formwork may represent as much as one-third of the total cost of a concrete structure, so the importance of the design and construction of this phase of a project cannot be overemphasized. The character of the structure, availability of equipment and form materials, anticipated repeated use of the forms, and familiarity with methods of construction influence design and planning of the formwork. Forms must be designed with a knowledge of the strength of the materials and the loads to be carried. The ultimate shape, dimensions, and surface finish must also be considered in the preliminary planning phase. Forms also provide protection for concrete, aid in curing, and support reinforcing rods and conduit which may be embedded within the concrete.

Use

Forms for concrete structures must be tight, rigid, and strong. If forms are not tight, there will be a loss of mortar which may result in honeycomb, or a loss or water that causes sand streaking. The forms must be braced enough to stay in alinement, and strong enough to hold the concrete. Special care should be taken in bracing and tying down forms, such as those for retaining walls, in which the mass of concrete is large at the bottom and tapers toward the top. In this type of construction and in other types, such as the first pour for walls and columns, the concrete tends to lift the form above its proper elevation. If the forms are to be used again they must be easily removed and re-erected without damage. Most forms are made of wood but steel forms are commonly used for work involving large unbroken surfaces, such as retaining walls, tunnels, pavements, curbs, and sidewalks. Steel forms for sidewalks, curbs, and pavements are especially advantageous since they can be used many times.

Materials

Forms are generally constructed from three different materials:

Earth. Earth forms are used in subsurface construction where the soil is stable enough to retain the desired shape of the concrete structure. The advantages of this type of form are that less excavation is required and there is better settling resistance. The obvious disadvantage is a rough surface finish, so the use of earth forms is generally restricted to footings and foundations.

Metal. Metal forms are used where added strength is required or where the construction will be duplicated at another location. Metal forms are more expensive, but they may be more economical than wooden forms if they can be used often enough. Examples of their use would be highway paving forms or curb and sidewalk forms.

Wood. Wooden forms are by far the most common type used in building construction. They have the advantage of economy, ease in handling, ease of production, and adaptability to many desired shapes. Added economy may result from reusing form lumber later for roofing, bracing, or similar purposes. Lumber should be straight, structurally sound, strong, and only partially seasoned. Kiln-dried timber has a tendency to swell when soaked with water from the concrete. If the boards are tight jointed the swelling causes bulging and distortion. If green lumber is used allowance should be made for shrinkage or the forms should be kept wet until the concrete is in place. Soft woods such as pine, fir, and spruce make the best and most economical form lumber since they are light, easy to work with, and available in almost every region. Lumber that comes in contact with the concrete should be surfaced on one side at least and on both edges. The surfaced side is turned toward the concrete. The edges of the lumber may be square, shiplap, or tongue and groove. The latter makes a more watertight joint and tends to prevent warping. Plywood can be used economically for wall and floor forms if it is made with waterproof glue and is identified for use in concrete forms. Plywood is more warp-resistant and can be reused more often than lumber. Plywood is made in thicknesses of $\frac{1}{4}$, $\frac{3}{8}$, $\frac{9}{16}$, $\frac{5}{8}$, and $\frac{3}{4}$ of an inch and in widths up to 48 inches. Although longer lengths are manufactured, 8-foot lengths are the most commonly used. The $\frac{5}{8}$-and $\frac{3}{4}$-inch thicknesses are most economical; the thinner sections will require solid backing to prevent deflection. The $\frac{1}{4}$-inch thickness is useful for curved surfaces.

Fiber Forms. Impregnated and waterproofed cardboard and other fiber materials are used as forms for round concrete columns and other appli-

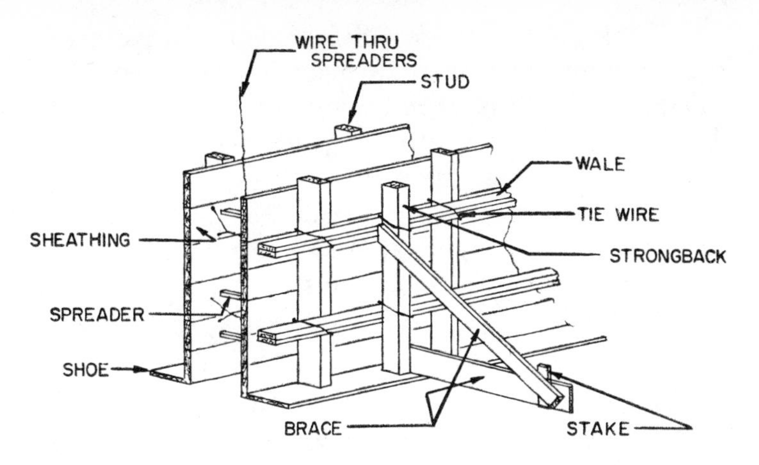

Figure 4–1. Form for a concrete wall.

cations where preformed shapes are desirable. These forms are usually made by gluing successive layers of fiber together and molding them to the desired shape. A major advantage of these forms is the time saved since form fabrication at the job site is not necessary.

Forming

Wall Forms. Elements of wooden forms for a concrete wall are shown in figure 4–1.

Sheathing. Sheathing forms the surfaces of the concrete. It should be as smooth as possible, especially if the finished surfaces are to be exposed. Since the concrete is in a plastic state when placed in the form, the sheathing should be watertight. Tongue and groove sheathing gives a smooth watertight surface. Plywood or hardboard can also be used.

Studs. The weight of the plastic concrete will cause the sheathing to bulge if it is not reinforced. Studs are run vertically to add rigidity to the wall form. Studs are generally made from 2 x 4 or 3 x 6 material.

Wales (Walers). Studs also require reinforcing when they extend over four or five feet. This reinforcing is supplied by *double wales*. Double wales also serve to tie prefabricated panels together and keep them in a straight line. They run horizontally and are lapped at the corners of the forms to add rigidity. Wales are usually made from the same material as the studs.

Braces. There are many types of braces which can be used to give the forms stability. The most common type is a diagonal member and horizontal member nailed to a stake and to a stud or wale. The diagonal member should make a 30° angle with the horizontal member. Additional bracing may be added to the form by placing vertical members (strongbacks) behind the wales or by placing vertical members in the corner formed by intersecting wales. Braces are not part of the form design and are not considered as providing any additional strength.

Shoe plates. The shoe plate is nailed into the foundation or footing and is carefully placed to maintain the correct wall dimension and alinement. The studs are tied into the shoe and spaced according to the correct design.

Spreaders. In order to maintain proper distance between forms, small pieces of wood are cut to the same length as the thickness of the wall and are placed between the forms. These are called spreaders. The spreaders are not nailed but are held in place by friction and must be removed before the concrete hardens. A wire should be securely attached to the spreaders so that they can be pulled out after the concrete has exerted enough pressure to the walls to allow them to be easily removed.

Tie wires. Tie wire is a tensile unit designed to hold the concrete forms secure against the lateral pressure of unhardened concrete. A double strand of tie wire is always used.

Column Forms. Elements of wooden forms for concrete columns are shown in figure 4–2.

Sheathing. In column forms, sheathing runs vertically to save on the number of sawcuts required. The corner joints should be firmly nailed to insure watertightness.

Batten. Batten are narrow strips of boards (cleats) that are placed directly over the joints to fasten the several pieces of vertical sheathing together.

Yokes. The horizontal dimensions on a column are small enough so that bracing is not

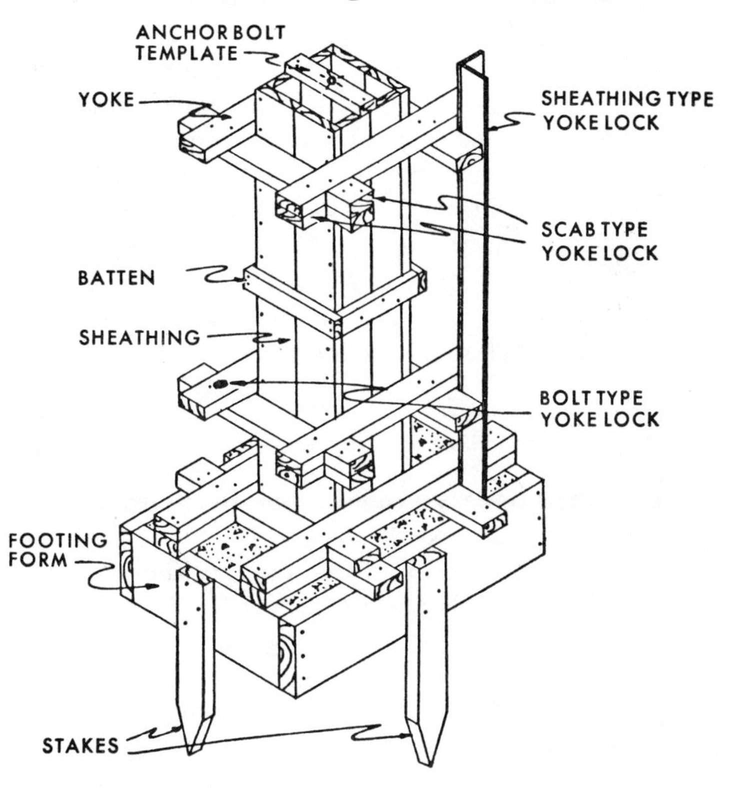

Figure 4–2. Form for a concrete column.

required in the vertical plane. A rectangular horizontal brace known as a yoke is used. The yoke wraps around the column and keeps the concrete from distorting the form. The yoke can be locked by the sheathing, scab, or bolt type yoke lock.

DESIGN

Strength Considerations

Forms for concrete construction must support the plastic concrete until it has hardened. Stiffness is an important feature in forms and failure to provide for this may cause unfortunate results. Forms must be designed for all the weight they are liable to be subjected to including the dead load of the forms, plastic concrete in the forms, the weight of workmen, weight of equipment and materials whose weight may be transferred to the forms, and the impact due to vibration. These factors vary with each project but none should be neglected. Ease of erection and removal are also important factors in the economical design of forms. Platform and ramp structures independent of formwork are sometimes preferred to avoid displacement of forms due to loading and impact shock from workmen and equipment.

Basis of Form Design

When concrete is placed in the form, it is in a plastic state and exerts hydrostatic pressure on the forms. The basis of form design, therefore, is the maximum pressure developed by the concrete during placing. The maximum pressure developed will depend on the rate of placing and the temperature. The rate of placing will affect the pressure because it determines how much hydrostatic head will be built up in the form. The hydrostatic head will continue to increase until the concrete takes its initial set, usually in about 90 minutes. However, at low temperatures, the initial set takes place much more slowly so it is necessary to consider the temperature, at the time of placing. Knowing these two factors, and the type of form material to be used, a tentative design may be calculated.

Wall Forms

Procedure. Is it desirable to design forms according to a step-by-step procedure to assure the consideration of all pertinent factors. Wooden forms for a concrete wall should be designed by the following steps:

Determine the materials available for sheathing, studs, wales, braces, shoe plates and tie wires.

Determine the mixer output by dividing the mixer yield by the batch time. Batch time includes loading all ingredients, mixing, and unloading. If more than one mixer will be used, multiply mixer output by the number of mixers.

$$\text{Mixer output (cu ft/hr)} = \frac{\text{Mixer yield (cu ft)}}{\text{Batch time (min)}} \times \frac{60 \text{ min}}{\text{hr}}$$

Determine the area that is enclosed by the forms.

$$\text{Plan area (sq ft)} = L \times W$$

Determine the rate (vertical feet per hour) of placing the concrete in the form by dividing the mixer output by the plan area.

$$\text{Rate of placing (ft/hr)} = \frac{\text{Mixer output (cu ft/hr)}}{\text{Plan area (sq ft)}}$$

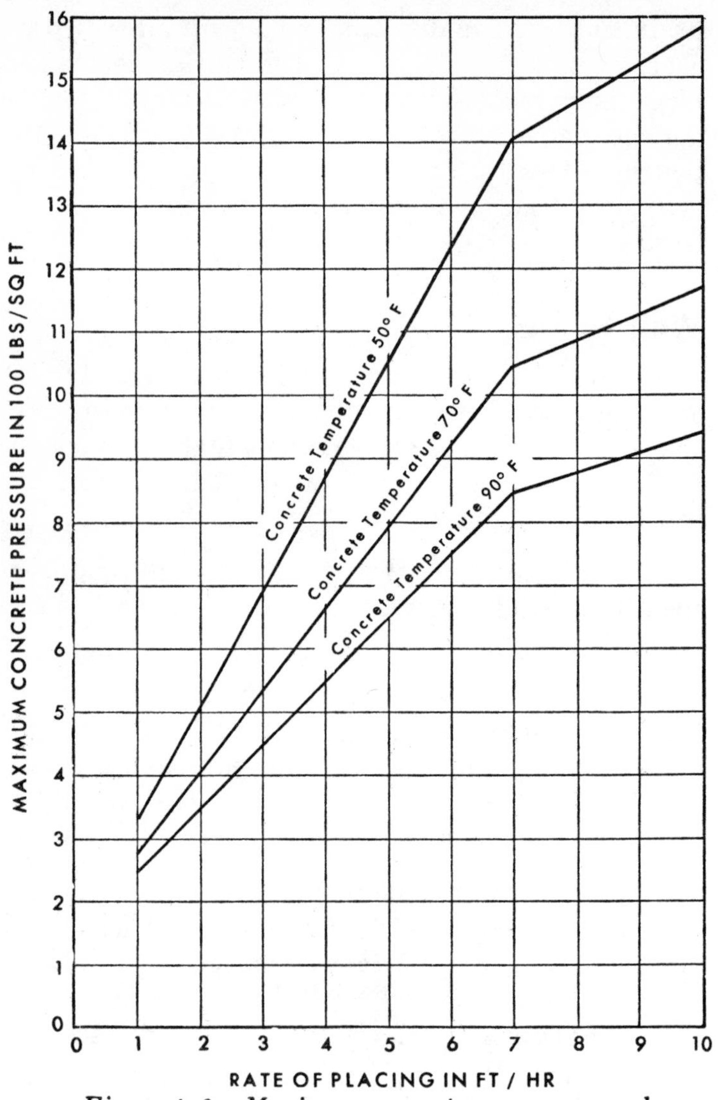

Figure 4–3. Maximum concrete pressure graph.

Make a reasonable estimate of the placing temperature of the concrete.

Determine the maximum concrete pressure by entering the bottom of figure 4–3 with the rate of placing. Draw a line vertically up until it intersects the correct concrete temperature curve. Read horizontally across from the point of intersection to the left side of the graph and determine the maximum concrete pressure.

Determine the maximum stud spacing by entering the bottom of figure 4–4 with the maximum concrete pressure. Draw a line vertically up until it intersects the correct sheathing curve. Read horizontally across from the point of intersection to the left side of the graph. If the stud spacing is not an even number of inches, round the value of the stud spacing down to the next lower even number of inches. For example, a stud spacing of 17.5 inches would be rounded down to 16 inches.

Determine uniform load on a stud by multiplying the maximum concrete pressure by the stud spacing.

Uniform load on stud (lb/lineal ft) =
Maximum concrete pressure (lb/sq ft) ×
stud spacing (ft)

Determine the maximum wale spacing by entering the bottom of figure 4–5 with the uniform load on a stud. Draw a line vertically up until it intersects the correct stud size curve. Read horizontally across from the point of intersection to the left side of the graph. If the wale spacing is not an even number of inches, round the value of the wale spacing down to the next lower even number of inches. Double wales (two similar members) are used in every case as shown in figure 4–1.

Determine the uniform load on a wale by multiplying the maximum concrete pressure by the wale spacing.

Uniform load on wale (lb/lineal ft) =
Maximum concrete pressure (lb/sq ft) × wale spacing (ft)

Determine the tie wire spacing, based on the wale size, by entering the bottom of figure 4–6 with the uniform load on a wale. Draw a line vertically up until it intersects the correct double wale size curve. Read horizontally across from the point of intersection to the left side of the graph. If the tie spacing is not an even number of inches, round the value of the tie spacing down to the next lower even number of inches.

Determine the tie wire spacing based on the tie wire strength by dividing the tie wire strength by the uniform load on a wale. If the tie wire spacing is not an even number of inches, round the computed value of the tie spacing down to the next lower even numer of inches. If possible, use a tie wire size that will provide a tie spacing equal to or greater than the stud spacing. Always use a double strand of wire. If the strength

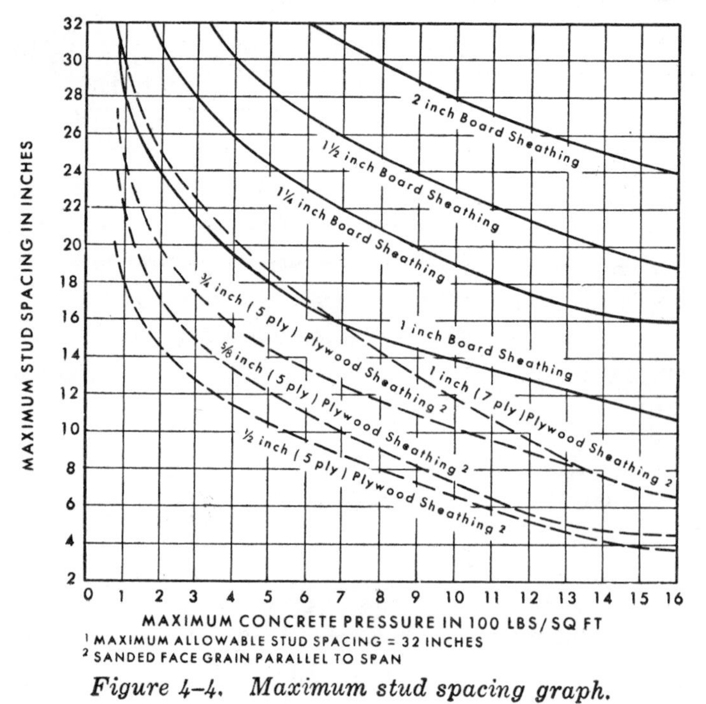

Figure 4–4. Maximum stud spacing graph.

of the available tie wire is unknown, the minimum breaking load for a double strand of wire (found in the Army supply system) is given in table 4–1.

$$\text{Tie wire spacing (in)} = \frac{\text{Tie wire strength (lbs)} \times \text{(12 in/ft)}}{\text{Uniform load on wale (lb/ft)}}$$

Determine the maximum tie spacing by selecting the smaller of the tie spacings based on the wale size and on the tie wire strength.

Compare the maximum tie spacing with the maximum stud spacing. If the maximum tie

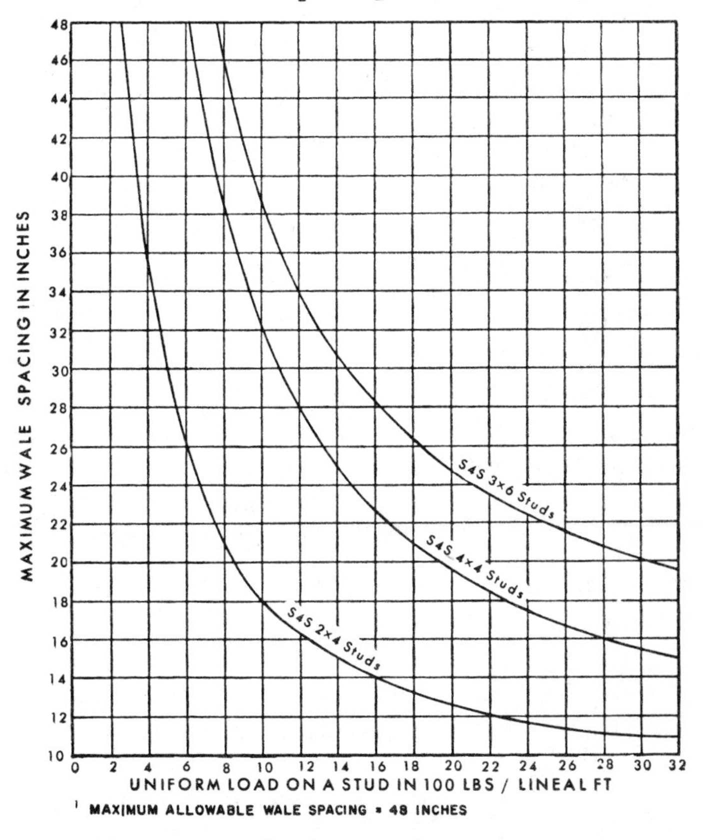

Figure 4–5. Maximum wale spacing graph.

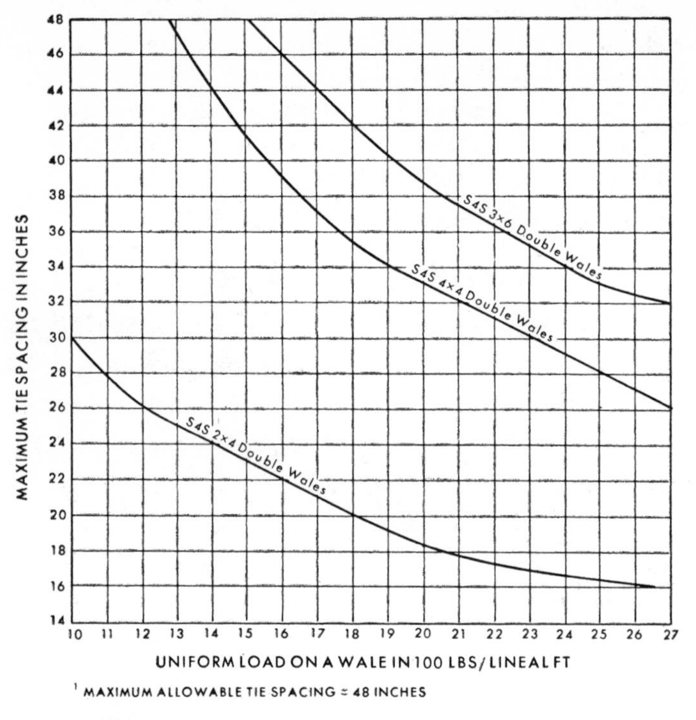

Figure 4–6. Maximum tie wire spacing.

spacing is less than the maximum stud spacing, reduce maximum stud spacing to equal the maximum tie spacing and tie at the intersections of the studs and wales. If the maximum tie spacing is greater than the maximum stud spacing, tie at the intersections of the studs and wales.

Determine the number of studs for one side of a form by dividing the form length by the stud spacing. Add one (1) to this number and round up the next integer. During form construction, place studs at the spacing determined above. The spacing between the last two studs may be less than the maximum allowable spacing.

$$\text{No. of studs} = \frac{\text{Length of form (ft)} \times 12 \text{ (in/ft)}}{\text{Stud spacing (in)}} + 1$$

Table 4-1. Breaking Load of Wire

STEEL WIRE

Size of wire Gage No.	*Minimum breaking load double strand Pounds*
8	1700
9	1420
10	1170
11	930

BARBED WIRE

Size of each wire Gage No.	*Minimum breaking load Pounds*
12½	950
13 [1]	660
13½	950
14	650
15½	850

[1] Single strand barbed wire.

Determine the number of wales for one side of the form by dividing the form height by the wale spacing and round up to the next integer. Place first wale one-half space up from the bottom and the remainder at the maximum wale spacing.

Determine the time required to place the concrete by dividing the height of the form by the rate of placing.

Example Form Design Problem: Design the forms for a concrete wall 40 feet long, 2 feet thick, and 10 feet high. A 16S Mixer is available and the crew can produce a batch (16 cu ft) of concrete every 5 minutes. The concrete temperature is estimated to be 70° F. Material available for use in constructing forms includes 2 x 4's and one-inch board sheathing.

Solution Steps:

1. Material available: 2 x 4's, one inch sheathing and No. 9 wire.

2. Mixer output $= \frac{16 \text{ cu ft}}{5 \text{ min}} \times \frac{60 \text{ min}}{\text{hr}} =$ 192 cu ft/hr

3. Plan area of forms $= 40 \text{ ft} \times 2 \text{ ft} =$ 80 sq ft

4. Rate of placing $= \frac{192 \text{ cu ft/hr}}{80 \text{ sq ft}} = 2.4$ ft/hr

5. Temperature of concrete: 70° F.

6. Maximum concrete pressure (fig. 4–3) = 460 lb/sq ft

7. Maximum stud spacing (fig. 4–4) = 18 +, use 18 inches

8. Uniform load on studs $= 460 \text{ lb/sq ft} \times \frac{18 \text{ in.}}{12 \text{ in/ft}} = 690 \text{ lb/ft}$

9. Maximum wale spacing (figure 4–5) = 23 + use 22 inches

10. Uniform load on wales $= 460 \text{ lb/sq ft} \times \frac{22 \text{ in}}{12 \text{ in/ft}} = 843 \text{ lb/ft}$

11. Tie wire spacing based on wale size (fig. 4–6) = > 30 inches

12. Tie wire spacing based on wire strength $= \frac{1420 \text{ lb} \times 12 \text{ in/ft}}{843 \text{ lb/ft}} = 20+$—use 20 inches

13. Maximum tie spacing = 20 inches

14. Maximum tie spacing is greater than maximum stud spacing, therefore, reduce the tie spacing to 18 inches and tie at the intersection of each stud and double wale

15. Number of studs per side $= (40 \text{ ft} \times \frac{12 \text{ in/ft})}{18 \text{ in}}) + 26.7 + 1$, use 28 studs

16. Number of double wales per side = $10 \text{ ft} \times \frac{12 \text{ in/ft}}{22 \text{ in}} = 5+$—use 6 double wales

17. Time required to place concrete = $\frac{10 \text{ ft}}{2.4 \text{ ft/hr}} = 4.17$ hrs.

Column Forms

Procedure. Wooden forms for a concrete column should be designed by the following steps:

Determine the materials available for sheathing, yokes, and battens. Standard materials for column forms are 2 x 4's and 1-inch sheathing.

Determine the height of the column.

Determine the largest cross-sectional dimension of column.

Determine the yoke spacings by entering table 4–2 and reading down the first column until the correct height of column is reached. Then read horizontally across the page to the column headed by the largest cross-sectional dimension. The center-to-center spacing of the second yoke above the base yoke will be equal to the value in the lowest interval that is partly contained in the column height line. All subsequent yoke spacings may be obtained by reading up this column to the top. This procedure gives maximum yoke spacings. Yokes may be placed closer together if desired. Table 4–2 is based upon use of 2 x 4's and 1-inch sheathing.

Example Problem: Determine the yoke spacing for a 9-foot column whose largest cross-sectional dimension is 36 inches. 2 x 4's and 1-inch sheathing are available.

Solution steps:

1. Material available—2 x 4's and 1-inch sheathing

Table 4–2. Column Yoke Spacing

LARGEST DIMENSION OF COLUMN IN INCHES – 'L'

HEIGHT: 1', 2', 3', 4', 5', 6', 7', 8', 9', 10', 11', 12', 13', 14', 15', 16', 17', 18', 19', 20'

L	Yoke spacings (from top of form downward)
16"	31", 31", 30", 29", 21", 20", 18", 15", 14", 13", 13", 12"
18"	29", 28", 28", 26", 20", 18", 16", 15", 13", 12", 12", 11", 11"
20"	27", 26", 26", 24", 19", 16", 15", 14", 12", 11", 11", 10", 10", 10", 9"
24"	23", 23", 23", 22", 16", 14", 13", 12", 11", 10", 9", 9", 9", 8", 8", 8"
28"	21", 21", 20", 18", 15", 13", 12", 10", 9", 9", 9", 8", 8", 8", 7", 6"
30"	20", 20", 19", 18", 18", 12", 12", 10", 9", 8", 8", 7", 7", 7", 6"
32"	19", 19", 18", 17", 13", 12", 10", 10", 8", 8", 7", 7", 6", 6"
36"	17", 17", 17", 15", 12", 11", 10", 8", 8", 7", 7", 6", 6", 6", 6"

L

2. Height of column is 9 feet

3. Largest cross-sectional dimension of the column is 36 inches.

4. Maximum yoke spacing for column (table 4–2) starting from the bottom of form are 8″, 8″, 10″, 11″, 12″, 15″, 17″, 17″ and 10″. The space between the top two yokes has been reduced because of the limits of the column height.

Foundation Forms

Foundation forms include forms for large footings, wall footings, column footings, and pier footings. These foundations or footings are relatively low in height and have a primary function of supporting a structure. The depth of concrete is usually small, therefore the pressure on the form is relatively low. Thus, design based on strength consideration generally is not necessary. Whenever possible, the earth should be evacuated and used as a mold for the concrete footings. Details for footing forms are given in paragraph 4–12.

Floor Forms

Procedure. Wooden forms for flat concrete slabs should be designed by the following steps:

Determine material available for sheathing, cleats, joists, stringers, and shores. Typical materials are—1-inch tongue and groove or ¾-inch plywood for sheathing; 1- by 4-inch cleats, 2- by 6-inch joists; 2- by 8-inch, 4- by 4-inch, or 4- by 6-inch stringers; and 4- by 4-inch shores.

Determine the total unit load on the floor form. The weight of ordinary concrete is assumed to be about 150 lb per cubic foot. Using this figure, the weight of concrete is 50 pounds per square foot for a 4-inch slab, 63 pounds per square foot for a 5-inch slab, and 75 pounds per square foot for a 6-inch slab. In addition, a live load for men and construction materials must be added. This is generally 50 pounds per square foot, however, 75 pounds per square foot is frequently used if powered concrete buggies are utilized.

Determine the spacing of floor joists. Table 4–3 indicates the joist spacing as a function of slab thickness and span of the joists for 2-by 4-inch joists. This table gives spans determined by considering joist strength only. It does not take into consideration the deflection of the sheathing. If this is of concern, it should be checked by separate calculation. The joist span may be shortened by the addition of stringers.

Determine the location of the stringers which support the joists. For short spans, it may not be necessary to use stringers.

Determine the spacing of the shores, or posts, which support the stringers. Maximum spans for stringers are given in table 4–4.

Example Problem. A 5-inch concrete floor slab has a span of 12 feet. Material available includes 1-inch tongue and groove sheathing, 1 x 4's, 2 x 6's, and 4 x 4's, and 2 x 8's. Determine the spacing of the joists, stringers, and shores.

Solution Steps:

1. Material available: 1-inch tongue and groove, 1 x 4's, 2 x 6's, 4 x 4's, and 2 x 8's.
2. Total unit load = live load + concrete load = 50 + 63 = 113 lb per sq ft
3. Spacing of joists (use 2 x 6's, table 4–3): Locate a stringer in the middle of the span, giving a joist span of 6 feet and a joist spacing of 4 feet.
4. Stringer spacing (use 2 x 8's) = 6 feet (from above).
5. Spacing of shores (use 4 x 4's, table 4–4) = 5 feet

*Table 4–3. Joint Spacing for 2- by 6-Inch Joist**

Concrete slab thickness	Joist span						
	4-foot	5-foot	6-foot	7-foot	8-foot	9-foot	10-foot
4″	4′ 0″	4′ 0″	4′ 0″	4′ 0″	4′ 0″	3′ 0″	2′ 6″
5″	4′ 0″	4′ 0″	4′ 0″	3′ 6″	3′ 0″	3′ 0″	2′ 6″
6″	4′ 0″	3′ 0″	2′ 6″	2′ 0″	2′ 0″	2′ 0″	2′ 0″

*Based only on joist strength, does not consider deflection of sheathing.

Table 4-4. Maximum Spans for Stringers

SLAB THICKNESS, in.	STRINGER SIZE	STRINGER SPACING*		
		5 ft	6 ft	7 ft
4	2×8	5′6″	5′	4′6″
	4×6	6′6″	6′	5′6″
5		5′6″	5′	4′6″
	4×6	6′	5′6″	5′
6	2×8	5′	4′5″	4′
	4×6	5′6″	5′	5′
7				4′6″
8		5′	4′6″	

*Spacing based on live load of 50 lb/sq ft. For live load of 75 lb/sq ft, increase slab thickness by 2 in. and use corresp. spacing.

Stair Forms

Various types of stair forms are used, including prefabricated forms. For moderate width stairs joining typical floors, design based on strength considerations is not generally necessary A typical wooden stairway form is discussed in paragraph 4-17.

CONSTRUCTION

Foundation and Footing Forms

Footing Forms. When possible, the earth should be excavated so as to form a mold for concrete wall footings. Otherwise, forms must be constructed. In most cases, footings for columns are square or rectangular. The four sides should be built and erected in panels. The earth must be thoroughly moistened before the concrete is placed. The panels for the opposite sides of the footing are made to exact footing width. The 1-inch-thick sheathing is nailed to vertical cleats spaced on 2-foot centers. See (a) in figure 4-7 which shows a typical form for a large footing. Two-inch dressed lumber should be used for the cleats and cleats spaced $2\frac{1}{2}$ inches from each end of the panel as shown. The other pair of panels (b), figure 4–7, have two end cleats on the inside spaced the length of the footing plus twice the sheathing thickness. The panels are held together by No. 8 or 9 soft black annealed iron wrapped around the center cleats. All reinforcing bars must be in place before the wire is installed. The holes on each side of the cleat permitting the wire to be wrapped around the cleat should be less than one-half inch in diameter to prevent leakage of mortar through the hole. The panels may be held in place with form nails until the tie wire is installed. All nails should be driven from the outside part way so that they may be withdrawn easily when the forms are removed. For forms 4 feet square or larger, stakes should be driven as shown. These stakes and 1 x 6 boards nailed across the top prevent spreading. The side panels may be higher than the required depth of footing since they can be marked on the inside to indicate the top of the footing. If the footings are less than 1 foot deep and 2 feet square, the forms can be constructed of 1-inch sheathing without cleats. Boards for the sides of the form are cut and nailed as shown in figure 4–8.

Footing and Pier Forms. It is often necessary to place a footing and a small pier at the same time. The form for this type of concrete construction is shown in figure 4–9. The units are similar to the one shown in figure 4–7. Support for the upper form must be provided in such a way that it does not interfere with the placement of concrete in the lower form. This is done by nailing 2 by 4's or 4 by 4's to the lower form as shown. The top form is then nailed to these support pieces.

Wall Footings. Form work for a wall footing is shown in figure 4–10 and methods of bracing the form are given in figure 4–11. The sides of the

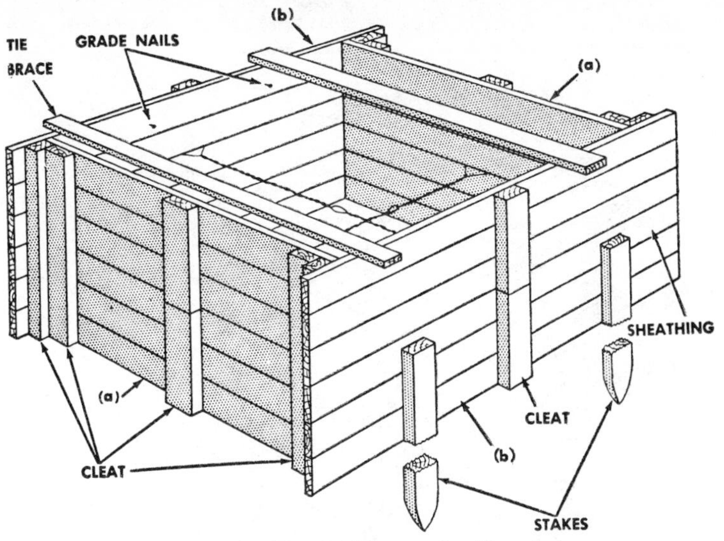

Figure 4–7. Typical large footing form.

forms are made of 2-inch lumber having a width equal to the depth of the footing. These pieces are held in place with stakes and are maintained the correct distance apart by spreaders. The short brace shown at each stake holds the form in line.

Wall Forms

Construction. Wall panels should be made up in lengths of about 10 feet so that they can be

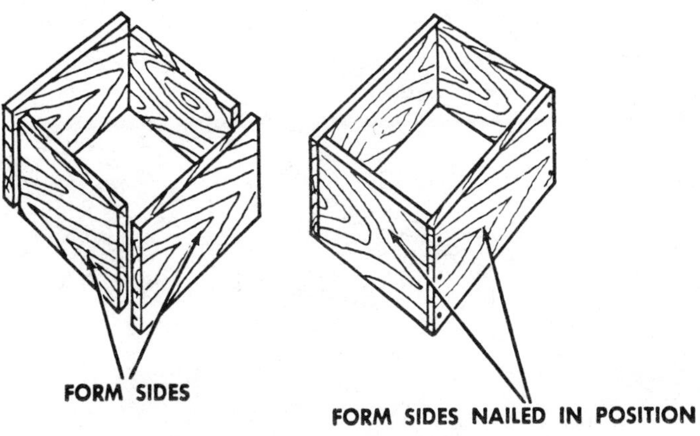

Figure 4–8. Small footing form.

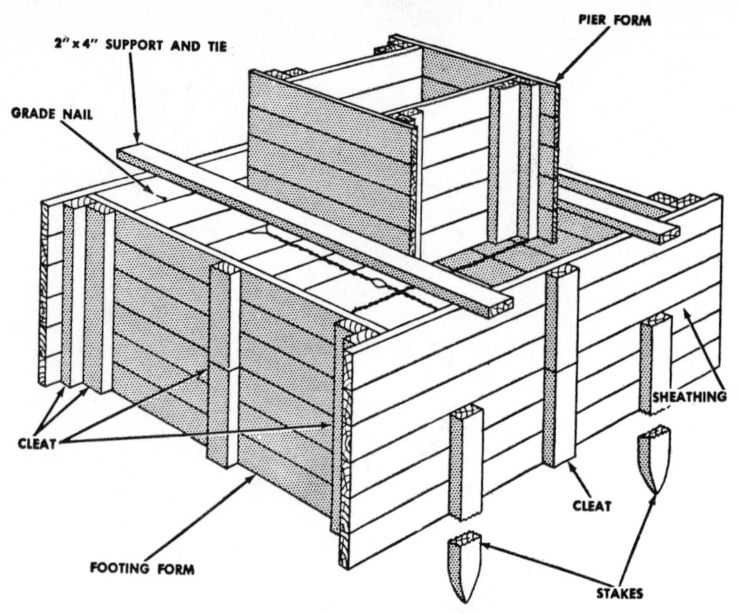

Figure 4–9. Footing and pier form.

easily handled. The panels are made by nailing the sheathing to the studs. Sheathing is normally 1-inch (13/16″ dressed) tongue and groove lumber, or ¾-inch plywood. The panels are connected together as shown in figure 4–12. Form details at the corner of a wall are given in figure 4–13.

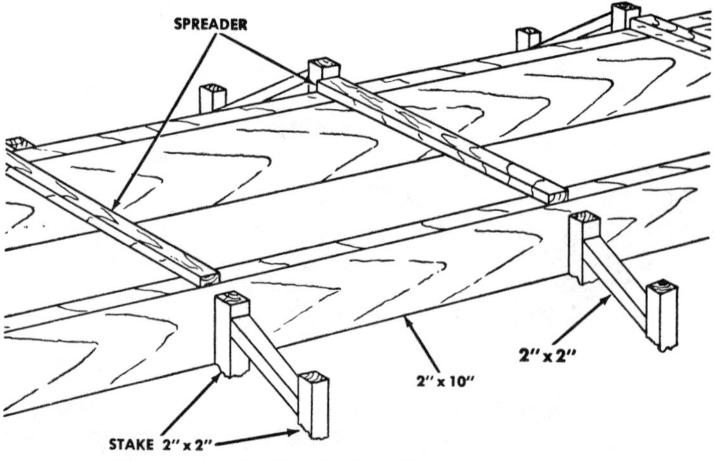

Figure 4–10. Typical wall footing form.

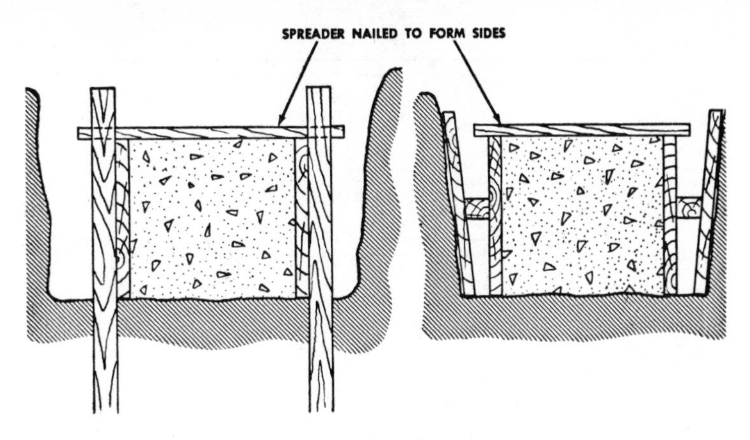

Figure 4–11. Methods of bracing wall footing forms.

Where curtain walls and columns are placed at the same time, the wall form construction should be as shown in figure 4–14. The wall form is made shorter than the clear distance between column forms to allow for a wooden strip to act as a wedge. In stripping the forms, the wedge is removed first to facilitate form removal.

Ties and Spreaders. The sides of the wall form must be kept apart with spreaders until the concrete is placed. They must also be tied together so they will not spread apart as the concrete is poured. Two methods of doing this with wire ties are shown in figures 4–15 and 4–16. The method shown in figure 4–15 illustrates the use of wire ties; this method should be used only for low walls

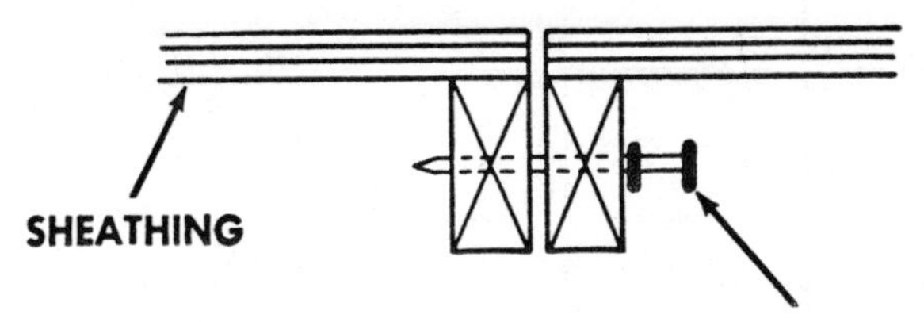

Figure 4–12. Method of connecting wall form panels together.

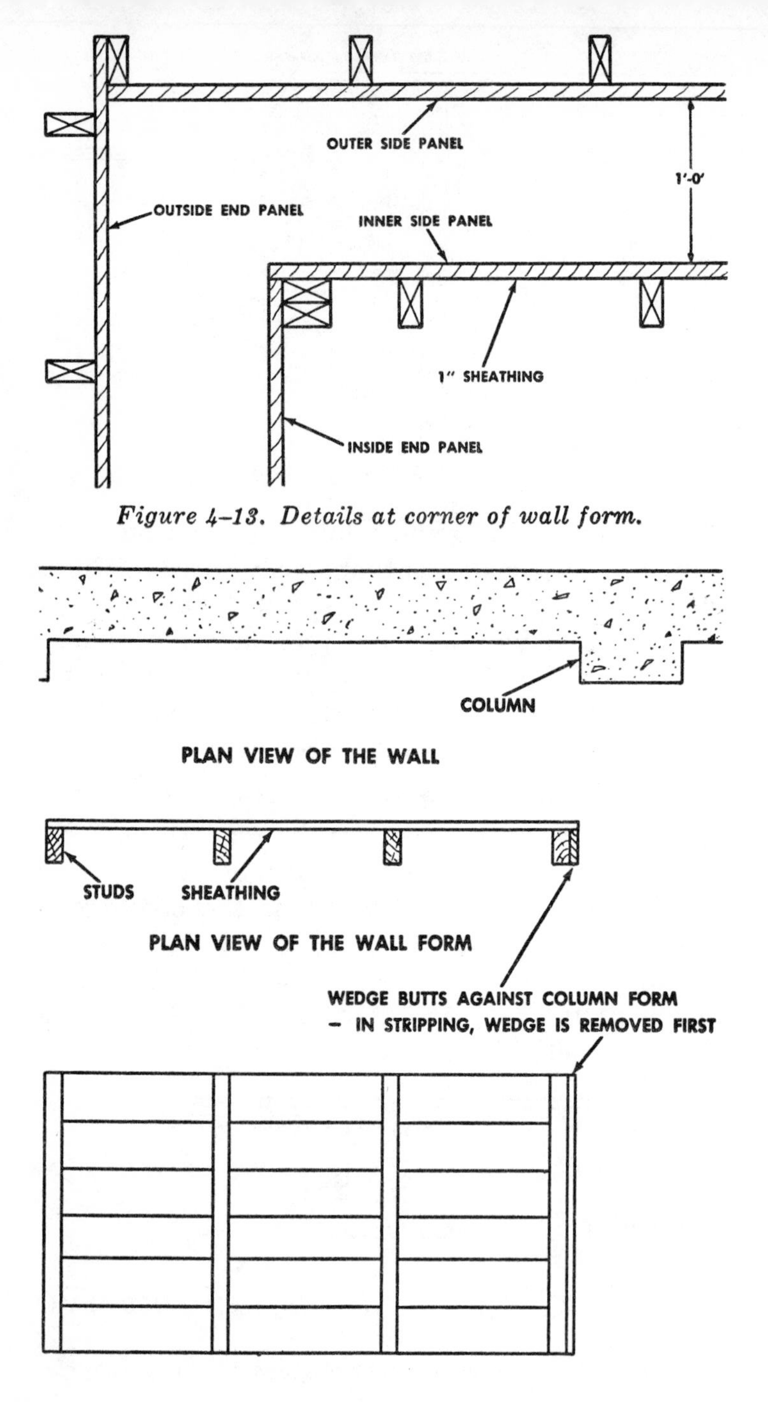

Figure 4–13. Details at corner of wall form.

Figure 4–14. Wall form for curtain walls.

spacing as the stud spacing but never more than 3 feet. A spreader should be placed near each tie wire. Each tie is formed by looping a wire around the wale on one form side. It is then brought through the forms, crossed inside the forms, and looped around the wale on the opposite side. The wire is finally made taut by twisting. When the wedge method is used the wire is first twisted around the point of the wedge. After this the nail is removed and the wedge driven tight in order to draw the wire taut. The wood spreaders shown

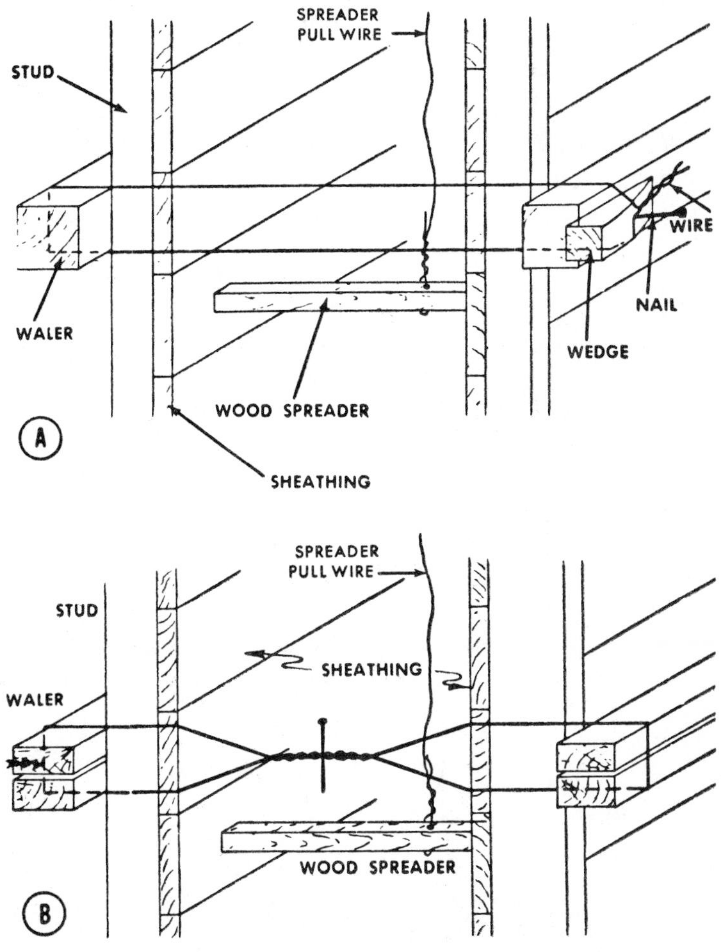

Figure 4-15. Wire ties for wall forms.

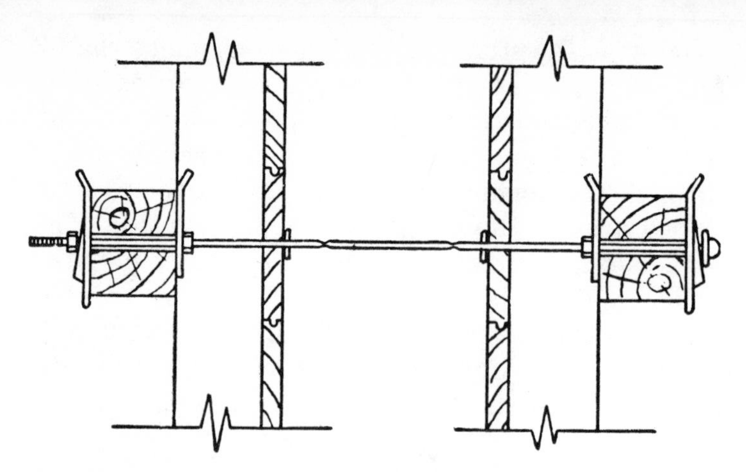

Figure 4–16. Tie rod and spreader for wall form.

and when tie rods are not available. The wire should be No. 8 or No. 9 gage soft black annealed iron wire, but in an emergency barbed wire can be used. The individual ties should have the same

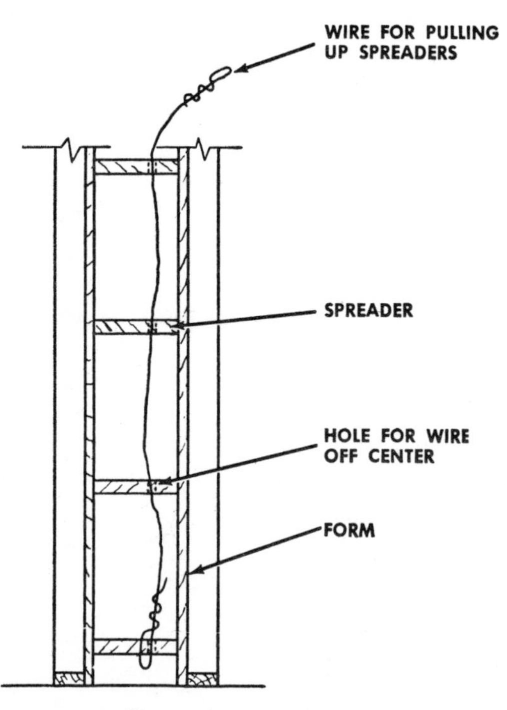

Figure 4–17. Removing wood spreaders.

must be removed as the forms are filled so that they will not become embedded in the concrete. A convenient way to remove these spreaders is shown in figure 4–17. A wire fastened to the bottom spreader is passed through holes drilled to one side of the center of each spreader. Pulling on the wire will remove the spreaders one after another as the concrete level rises in the forms. The tie rod and spreader illustrated in figure 4–16 serves both as a tie and a spreader. It should have the same spacing as tie wires. After the form is stripped, the rod is broken off at the notch which is slightly inside the concrete surface. If rust stains are objectionable and if the wall forms are to be maintained in exact position, this type tie and spreader should be used. If appearance is important, the holes made by breaking off the tie bar should be grouted in with a mortar mix.

Beam and Girder Forms

Beam Form. The type of construction to be used for beam forms depends upon whether the form is to be removed in one piece or whether the sides are to be stripped and the bottom left in place until such time as the concrete has developed enough strength to permit removal of the shoring. The latter type beam form is preferred; details for this type are shown in figure 4–18. Beam forms are subjected to very little bursting pressure but must be shored up at frequent intervals to prevent sagging under the weight of the fresh concrete.

Beam Form Constructions. The bottom of the form has the same width as the beam and is in one piece the full width. The sides of the form should be 1-inch thick tongue and groove sheathing and should lap over the bottom as shown in figure

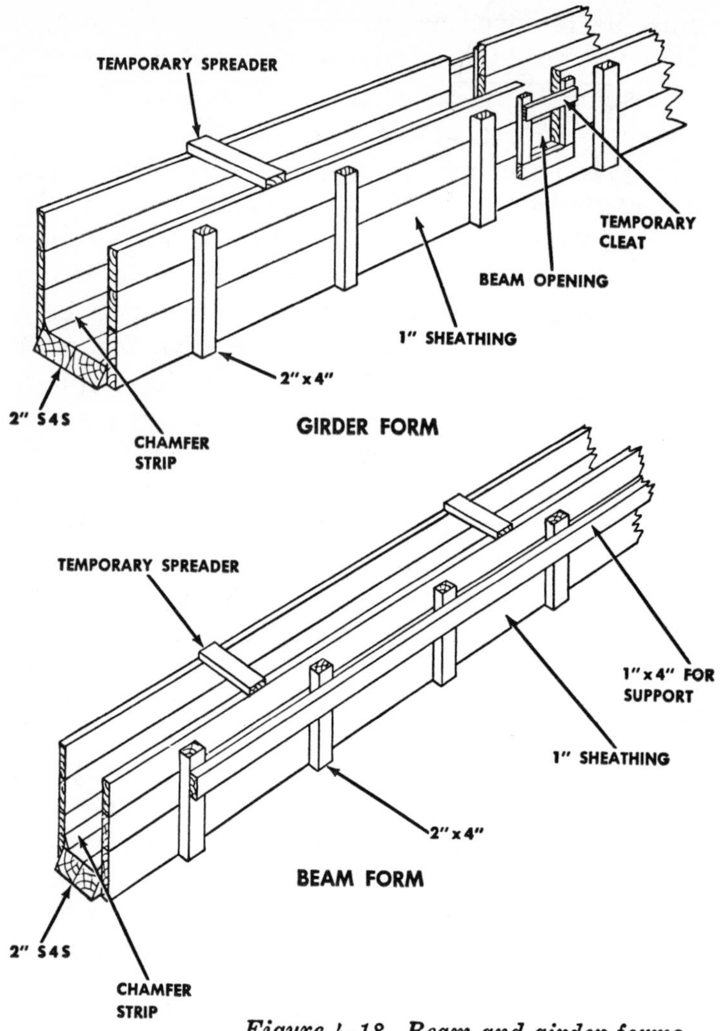

Figure 4–18. Beam and girder forms.

4–19. The sheathing is nailed to 2- by 4-inch struts placed on 3-foot centers. A 1- by 4-inch piece is nailed along the struts. These pieces support the joist for the floor panel, as shown. The sides of the form are not nailed to the bottom but are held in position by continuous strips as shown in detail E.

The cross pieces nailed on top serve as spreaders. After erection, the slab panel joints hold the beam

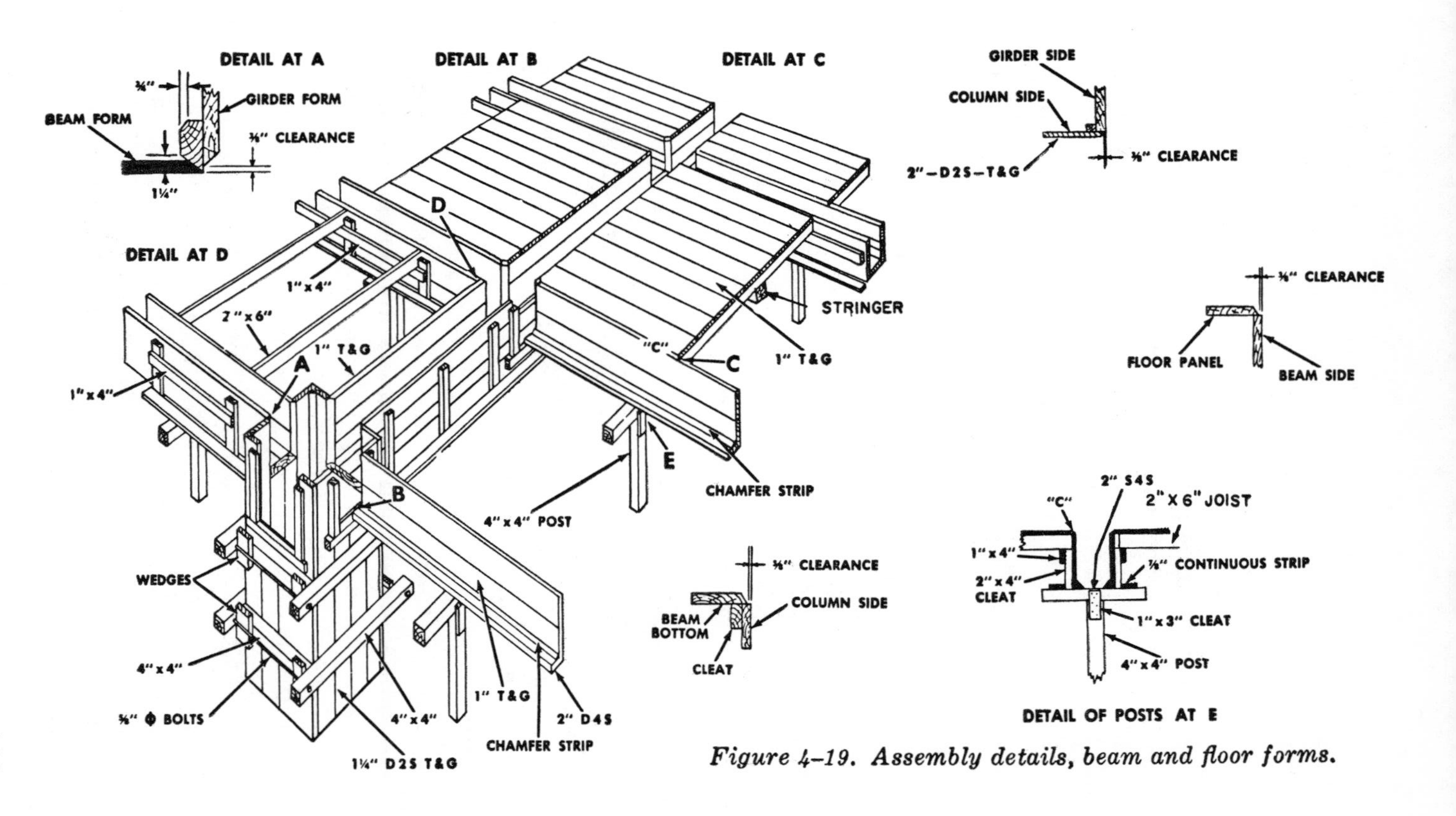

Figure 4–19. Assembly details, beam and floor forms.

sides in position. Girder forms are the same as beam forms except that the sides are notched to receive the beam forms. Temporary cleats should be nailed across the beam opening when the girder form is being handled.

Assembly. The method of assembling beam and girder forms is illustrated in figure 4–19. The connection of the beam and girder is illustrated in Detail D. The beam bottom butts up tightly against the side of the girder form and rests on a 2- by 4-inch cleat nailed to the girder side. Detail C shows the joint between beam and slab panel and details A and B show the joint between girder and column. The clearances given in these details are needed for stripping and also to allow for movement that will occur due to the weight of the fresh concrete. The 4 by 4 posts used for shoring the beams and girders should be spaced so as to provide support for the concrete and forms and wedged at either the bottom or top for easy dismantling.

Column and Footing Forms

Column Forms. Forms for square columns are made of wood and forms for round columns are made of steel or cardboard impregnated with a waterproofing compound.

Construction. Figure 4–2 shows an assembled column and footing form. The forms for the column sides are built in units and nailed to the yokes. The yoke spacing will be determined from table 4–2 according to instructions contained in paragraph 4–8.

Erection. The column form is erected as a unit after the steel reinforcing is assembled and tied to dowels in the footing. Placement of reinforcing steel in columns is described in chapter 6. The form should have a clean-out hole in the bottom through which debris can be removed. The

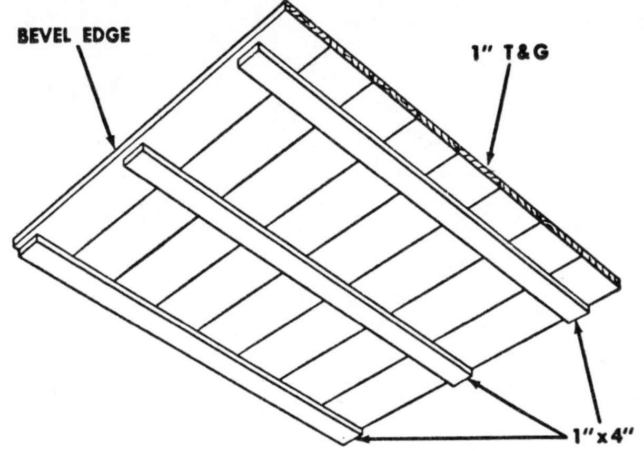

Figure 4–20. Form for floor slab.

pieces of lumber removed to make the clean-out holes should be nailed to the form so that they may be put back in place before concrete is placed in the column.

Floor Forms

The floor panels referred to in paragraph 4–14 for the system are constructed as shown in figure 4–20. The 1-inch-thick tongue and groove sheathing is nailed to 1- by 4-inch cleats on 3-foot centers. Three-quarter-inch plywood may be substituted for the 1-inch-thick tongue and groove sheathing. These panels are supported by 2- by 6-inch joists. The spacing for these joists is obtained from table 4–3. The spacing of the joists depends upon the thickness of the concrete slab to be supported and the span of the joist. If the concrete floor slab spans the distance between two walls, the panels are used in the same manner as when beams support the floor slab.

Stair Forms

Figure 4–21 shows one method of constructing stair forms for widths up to and including 3 feet. The sloping wooden platform that makes up the form for the underside of the steps should be made of 1-inch-thick tongue and groove sheathing.

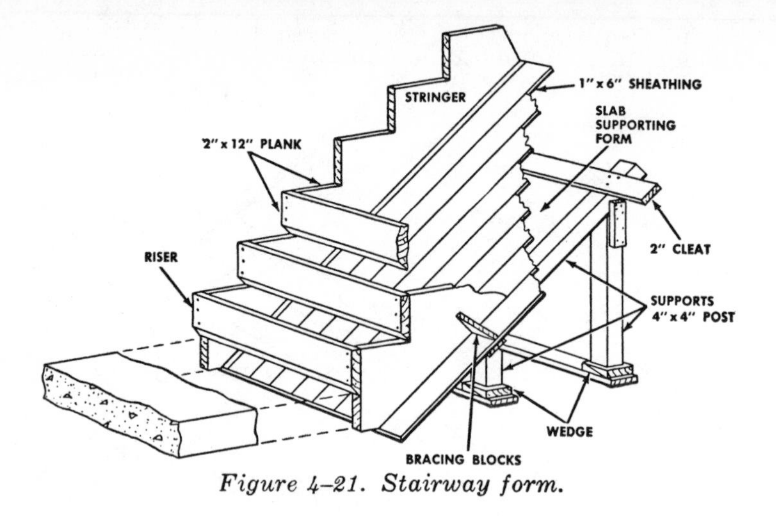

Figure 4–21. Stairway form.

This panel should extend about 12 inches beyond each side of the stairs to provide a support for the stringer bracing blocks. The back of the sloping panel should be shored up with 4 by 4's as shown. The 2- by 6-inch cleats nailed to the shoring should be spaced on 4-foot centers. The 4- by 4-inch shoring should rest on wedges to permit easy adjustment and to make it possible to remove the posts without any difficulty. The side stringers are made by 2- by 12-inch lumber cut as required for the tread and risers. The piece forming the riser should also be 2-inch-thick material beveled as shown.

Steel Pavement Forms

Standard steel pavement forms are rapidly and accurately set and should be used when obtainable. They are made of steel plate in sections 10 feet long. They provide a traction surface for form riding equipment and contain the concrete. Forms for slab thicknesses of 6, 7, 8, 9, 10, and 12 inches can normally be obtained. The usual base plate width is 8 inches. Locking plates rigidly join the ends of the sections. Three steel stakes per 10-foot section of form are required. These stakes are 1 inch in diameter.

After the stakes are driven, the form is set true to line and grade and clamped to the stakes by special splines and wedges. Forms are set on the subgrade ahead of the paver and must be left in place for at least 24 hours after the concrete is placed.

The forms must be oiled immediately before the concrete is deposited. After the steel form is completely oiled, the top side or traction edge of the form must be wiped free of oil.

The forms are removed as soon as practicable, cleaned, oiled, and stacked in neat piles ready for reuse.

Oiling and Wetting Forms

Oil for Wood Forms. Before concrete is placed in wood forms, the forms must be properly treated with a suitable form oil or other coating material which will prevent the concrete from sticking. Oil used on wood forms should penetrate the wood and prevent absorption of water. Almost any light-bodied petroleum oil will meet these qualifications. Sealing compounds are sometimes used to preserve the form material. Shellac applied to plywood is more effective than oil in preventing moisture from raising the grain and detracting from the finished surface of the concrete. Several commercial lacquers and similar products are also available for this purpose. If wooden forms are to be reused repeatedly, a coat of paint will help to preserve the wood. Occasionally, lumber delivered to the job will contain sufficient tannin or other organic substance to cause softening of the surface concrete. When this condition is recognized it can be remedied by treating the form surface with whitewash or limewater before applying the form oil or coating.

Oil for Steel Forms. Column and wall forms should be oiled before erection. All other forms can be oiled when most convenient but must be oiled before reinforcing steel is in place. Oils used on wood forms may not be suitable for steel forms; specially compounded petroleum oils are satisfactory. Synthetic castor oil and some marine engine oils are typical examples of compounded oils which have given good results on steel forms.

Application of Oil. Successful use of form oil depends on the manner of application and the condition of the form. Forms should be clean and surfaces should be smooth. Cleaning with wire brushes may mar the surface and cause sticking. The oil or coating should be applied by brush, spray, or swab and should cover the forms evenly and not be permitted to get on construction joint surfaces or reinforcing bars. Excess oil should be removed. The oil or other form coating used should not soften or stain the concrete surface or prevent the wet surfaces from being water-cured or hinder the proper functioning of sealing compounds used for curing.

Other Coating Materials. Fuel oil is a satisfactory coating material except during warm weather. One part petroleum grease to three parts of fuel oil will provide enough thickness for warm weather. Asphalt paints, varnish, and boiled linseed oil are also suitable coating.

Wetting Forms. If form oil is not obtainable, wetting may be substituted to prevent sticking but this should be as an emergency measure only.

Safety Precautions

Form Construction.

Protruding nails are one of the principal sources of accidents on form work.

Tools, particularly hammers, should be inspected frequently.

Mud sills should be placed under shoring that rests in the ground.

Precautions should be taken when work is progressing on elevated forms to protect both men on the scaffolds and personnel on the ground.

Raising of large form panels should not be attempted in heavy gusts of wind either by hand or by crane.

All shoring should be securely braced to preclude possible collapse of formwork.

b. Stripping Forms.

(1) Only workmen actually engaged in stripping the forms should be permitted in the immediate work area.

(2) No forms should be removed until the concrete has set.

(3) Stripped forms should be piled immediately to avoid congestion, exposed nails, and other hazards.

(4) Wires under tension should be cut with caution to avoid backlash.

JOINTS AND ANCHORS

Construction Joints

Purpose. Construction joints are used between the units of a structure and located so they will not cause weakness. It would be preferable theoretically that each beam, girder, column, wall, or floor slab be placed in one operation to produce a homogeneous member without seams or joints, but for practical reasons, this procedure is usually impossible. The planes separating the work done on different days, called construction joints, are placed where they will cause the minimum amount of weakness to the structure, and where the shearing stresses and bending moments are small or where the joints will be supported by other members.

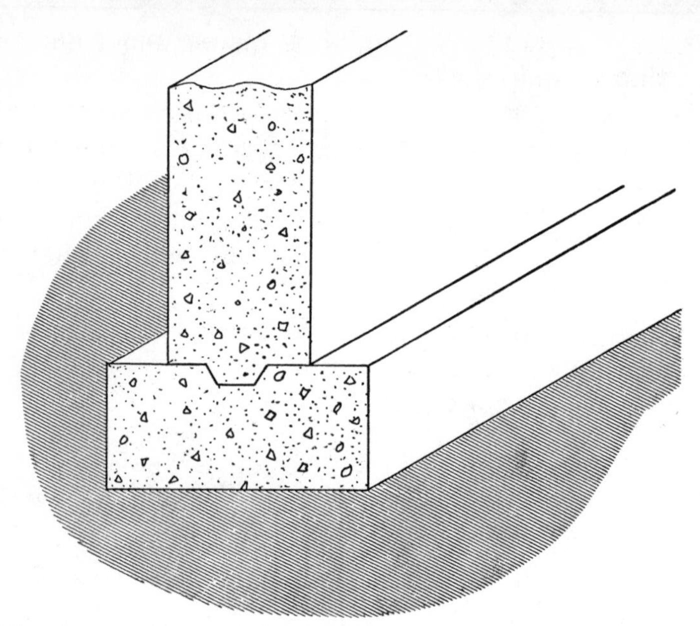

Figure 4–22. Construction joint between wall and footing.

Construction Joints Between Wall and Footing. At a construction joint between a wall and its footing, a keyway is usually necessary to transfer the shearing stresses. A keyway must always be used if no reinforcing steel or dowels tie the wall and footing together. This keyway can be formed by pressing a slightly beveled 2 by 4 into the concrete before it has set and removing the 2 by 4 after the concrete has hardened. The 2 by 4 should be well oiled before it is used. Such a keyway is shown in figure 4–22. If the wall and footing can be placed at one time, a construction joint is not necessary.

Vertical Construction Joints. If it is desirable to deposit the concrete for the full wall height, the forms should be divided into sections by vertical bulkheads as shown in figure 4–23. These bulkheads should be spaced so that the complete section can be filled in one continuous operation. Experience has shown that the V-joint in figure 4–24

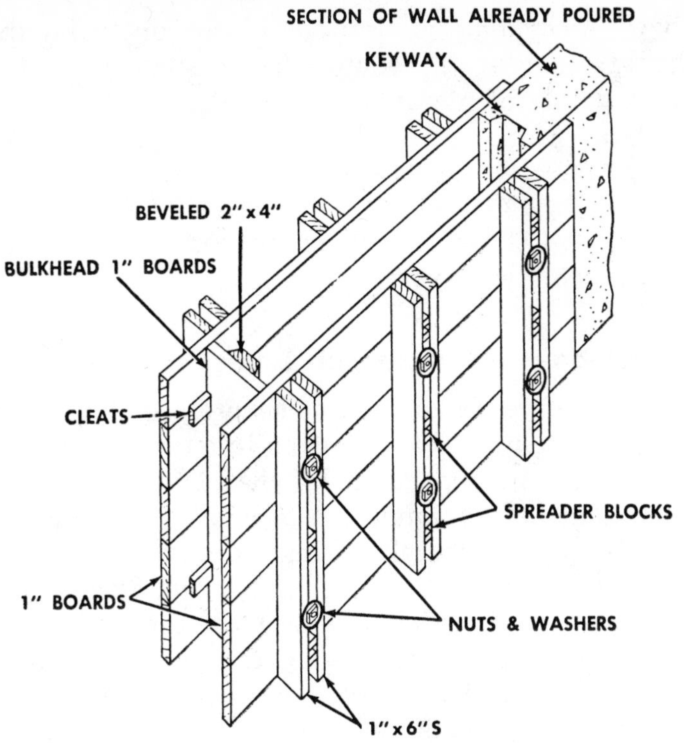

Figure 4–23. Vertical bulkhead in wall.

is less likely to break off than the joint shown in figure 4–23. If reinforcing steel or dowels cross the joint no projection is needed.

Beam, Column, and Floor Slab Joints. The proper place to end a pour in construction involving beams, columns, and floor slabs is shown in figure 4–25. The concrete in each column should be placed to the underside of the beam or floor slab

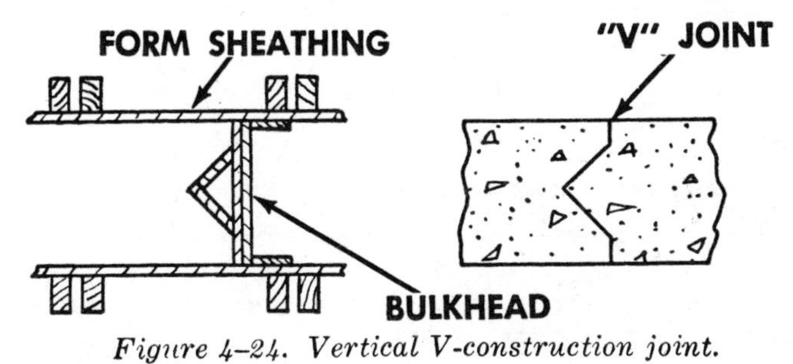

Figure 4–24. Vertical V-construction joint.

above. A construction joint in a beam or floor slab should occur at the center of the span so as to avoid points of maximum shear. All construction joints in beams and slabs should be vertical. Reinforcing steel or dowels should extend across the joint. A beam or slab should never be placed in two lifts vertically, for this produces a weak joint between the two layers.

Expansion and Contraction Joints

Principle. Shrinkage of concrete during hydration is comparable to a drop in temperature of 30° to 80° F. depending on the richness of the mix. Contraction joints are necessary to permit the concrete to shrink during the curing process without damage to the structure. The structural engineer will consider the best place for the joints from a standpoint of serving the purpose. They are usually placed where there is a change in thickness, at offsets and where the concrete will

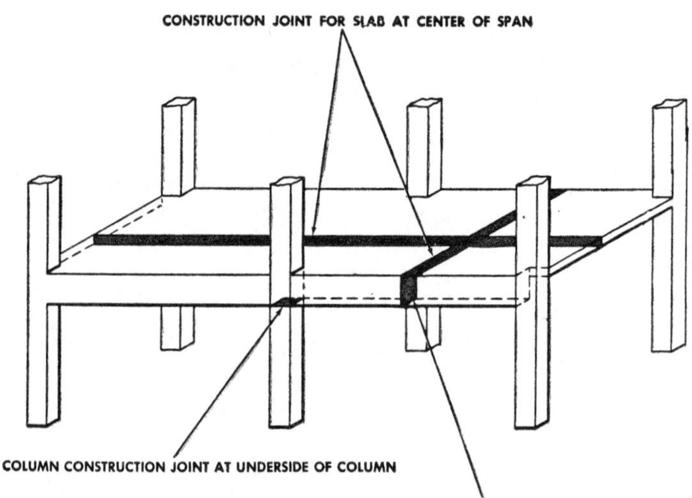

Figure 4–25. Location of construction joints in beams, columns, and floor slabs.

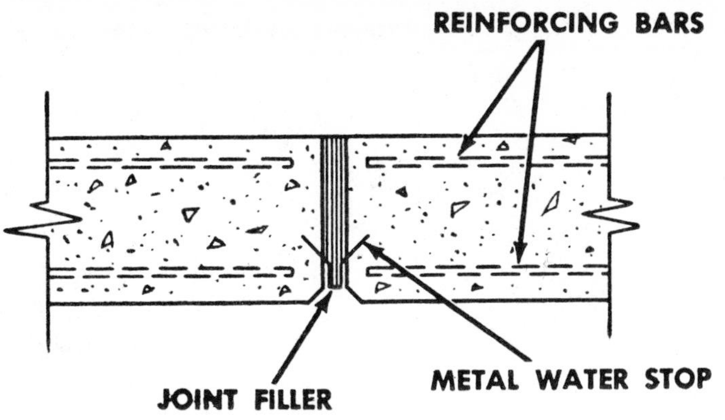

Figure 4–26. Expansion joint for wall.

tend to crack if shrinkage and deformations due to temperature are restrained. Joints should be about 30 feet center-to-center in exposed structures.

Expansion Joints. Expansion joints, in the form of vertical joints through the concrete, are used to isolate adjacent units of a structure, to prevent cracking due to shrinkage or temperature changes. Generally, an expansion joint is used with a premolded mastic or cork filler, if an elongation of adjacent parts and a closing of the joint is anticipated. Expansion joints for different types of structures are illustrated in figures 4–26 through 4–28. Expansion joints should be installed every 200 feet.

Contraction Joints. The purpose of contraction joints is to control cracking due to temperature changes incident to shrinkage of the concrete. If the concrete cracks it will usually occur at these joints. Usually, the contraction caused by shrinkage will offset a large part of the expansion due to a rise in temperature. Contraction joints are usually made with no filler or with a thin paint coat of asphalt, paraffin, or some other material to break the bond. Joints as shown in figure 4–29 should be installed at 30-foot intervals or closer

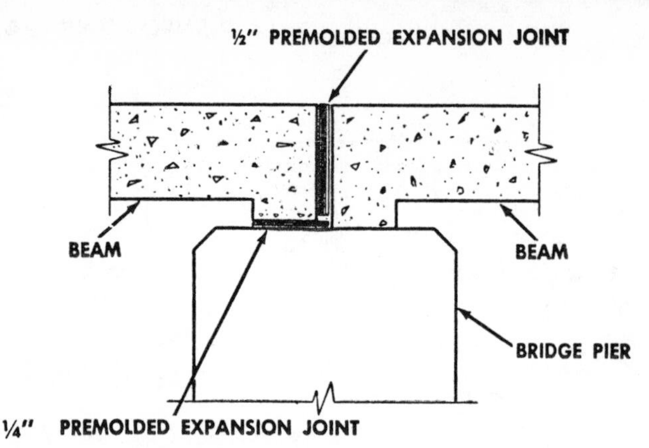

Figure 4–27. Expansion joint for bridge.

depending on the extent of local temperature change. These dummy contraction joints are formed by cutting to a depth of one-fourth to one-third the thickness of the section. Contraction and expansion joints are not used in beams and columns. Contraction and expansion joints in rein-

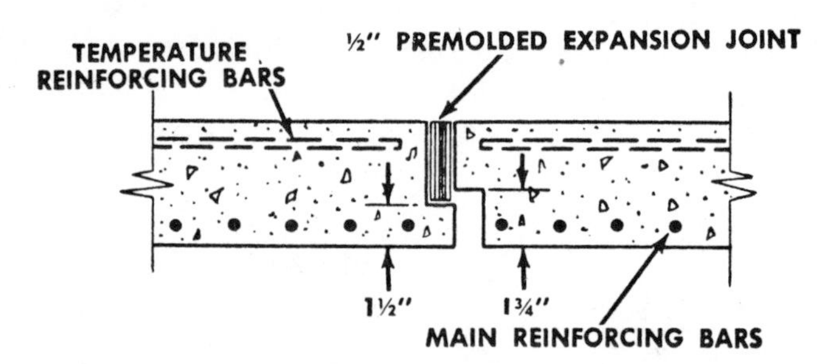

Figure 4–28. Expansion joint for floor slab.

forced concrete floor slabs should be placed at 100-foot intervals in each direction.

d. Concrete Pavement Joints. Concrete pavement joints are shown in figure 4–30.

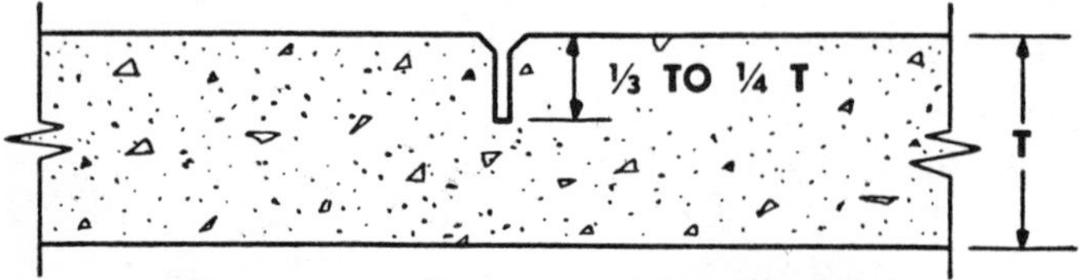

Figure 4–29. Dummy contraction joint.

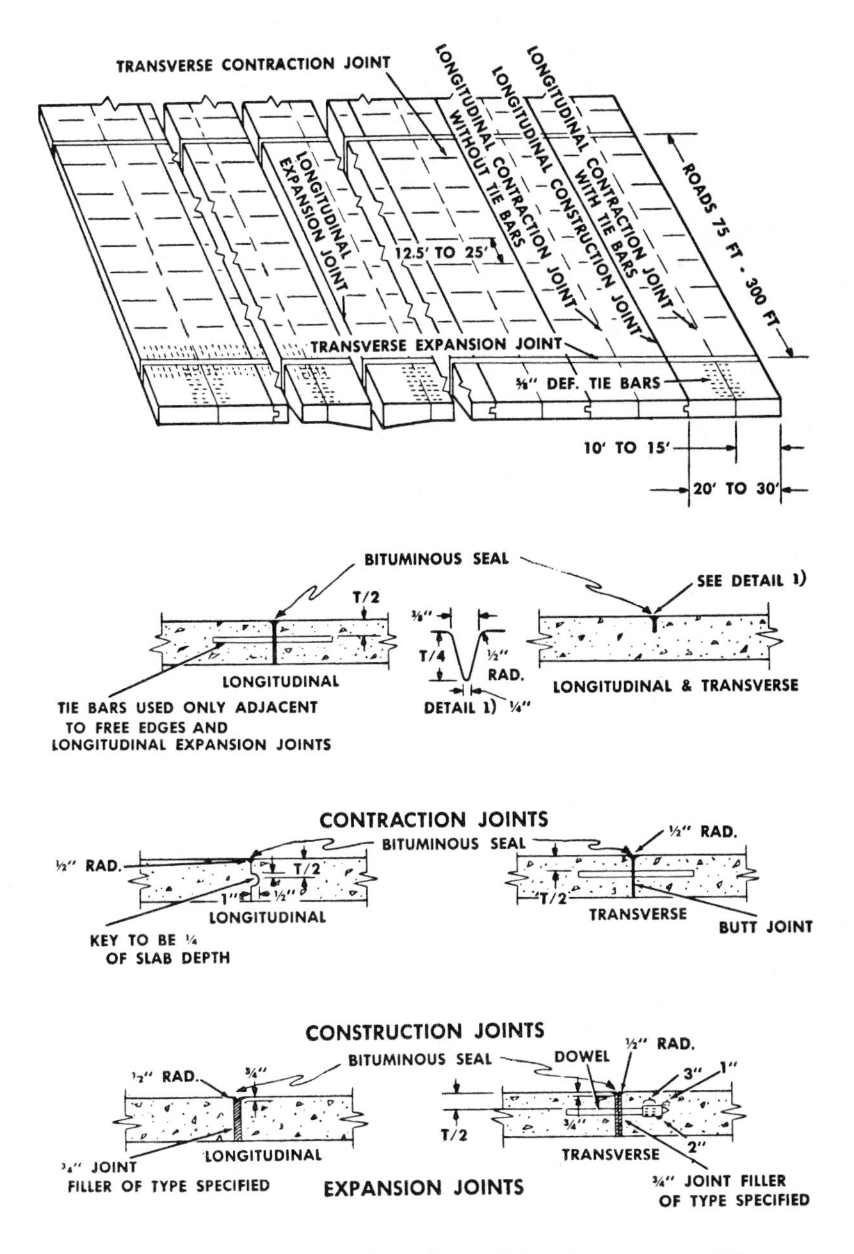

Figure 4–30. Design and spacing of joints for concrete slab.

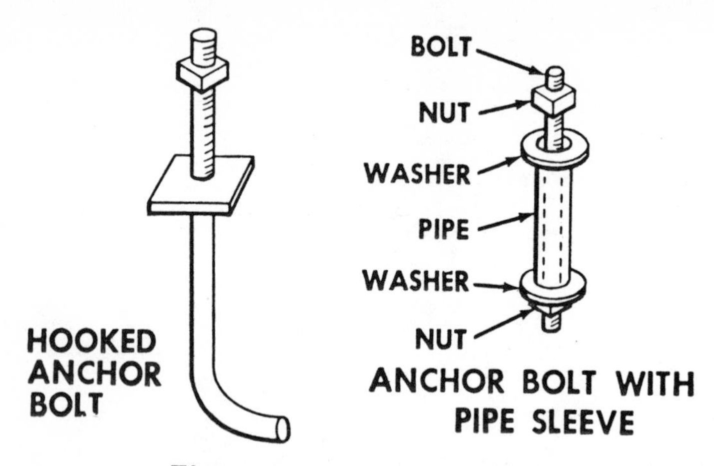

Figure 4–31. Anchor bolt details.

Anchor Bolts

Anchor Bolt with Pipe Sleeve. Two types of anchor bolts are shown in figure 4–31. The bolt and pipe sleeve is used to anchor machinery or structural steel. The sleeve which should be at least 1 inch larger in diameter than the bolt, permits the bolt to be shifted to compensate for any small error that may occur in positioning. The sleeves should be packed to prevent concrete from entering during the paving operation. When the machinery is set on a raised base 4 or more inches above the floor the pipe sleeve is not used. In this

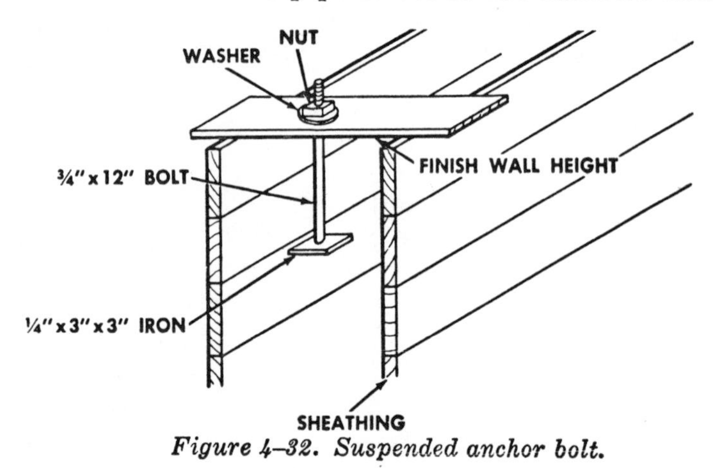

Figure 4–32. Suspended anchor bolt.

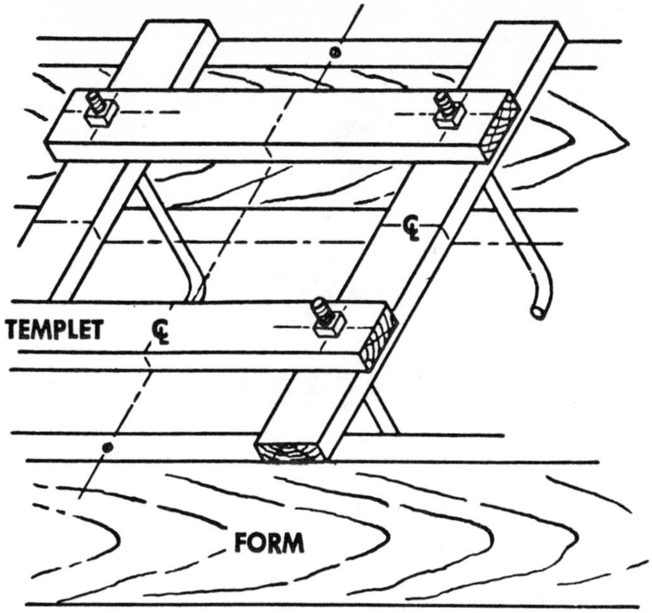

Figure 4–33. Anchor bolts held in place by a templet.

case, the anchor bolts are set in the floor so as to extend above the base. The machine is set, blocked up on scrap steel and leveled and the anchor bolts tightened. Then the concrete base is poured by packing the concrete into place from the sides.

Hooked Anchor Bolt. The hooked type of anchor bolt (fig. 4–31) is used to bolt a wood sill to a concrete or masonry wall. The bolt shown in figure 4–32 is also used for this purpose. Care must be taken in placing concrete to avoid disturbing the alinement of suspended bolts. The holes in the board from which the anchor bolt is suspended should be $\frac{1}{16}$ inch larger than the bolt to permit adjustment of the bolt.

Use of Templet. A templet is commonly used as shown in figure 4–33 to hold anchor bolts in place while the concrete is being poured. If sleeves are used they should be adjusted in the templet so that the top of the sleeve will be level with the top of the finished concrete.

5

Construction Procedures

A thorough and efficient reconnaissance of the construction site is done as the first step in the construction procedure. At this time any anticipated problems in the clearing and draining of the site and in the transporting and storing of materials are noted. In addition the site should be investigated for any unusual aspects, such as an undesirable soil or rock base, which may present construction problems. The anticipation and prior consideration of such problems will aid in avoiding unnecessary delays in construction.

Selection of Route. The selection of the best route to the construction site is based on the local traffic patterns, quality of existing roads and bridges, and the equipment to be used. Maximum use shall be made of the existing road net, since time and effort can generally be saved in repairing or improving an existing road rather than constructing a new one. When possible, an alternate route should be designated.

Location of Water and Aggregate. The nearest or most convenient source of suitable mixing water must be determined. In some instances, alternate sources may be noted in the event that subsequent tests indicate that the first choice is not suitable. Whenever possible local sources of sand and gravel will be used. These sources should be located and any necessary tests designated.

Estimation of Time. A carefully considered estimation of the time required for site preparation will be made, based on the reconnaissance of the area. A good estimate assures that the proper equipment is available where needed and when needed.

SITE PREPARATION

Most new construction takes place on land which has not been previously developed. In order that materials may be delivered to the site, approach roads must be built. Although these roads may have only temporary utility, they must be carefully constructed because of the heavy loads to which they are subjected. It is also possible that the routes will be used for permanent roads. There should be enough lanes constructed so that a free flow of traffic can reach the construction site.

Clearing and Draining of Site

Land Clearing. Land clearing is a construction operation consisting of clearing the construction site of all trees, downed timber, brush, other vegetation and rubbish; removing surface boulders and other material embedded in the ground; and disposing of all material cleared. Heavy equipment may be necessary to clear the site of large timbers and boulders. Hand equipment, explosives, and fire also have applications in site clearing. The factors determining the methods to be used are the acreage to be cleared, the type and density of vegetation, the terrain as it affects the operation of equipment, the availability of equipment and personnel, and the time available for completion.

Drainage. Adequate drainage must be provided in areas which have high ground-water tables and to carry off rain during the actual construction. Surface and sub-surface water can be withdrawn from the building site by using a well-point system or mechanical pumps.

Location of Building Site

The building site can be staked out after the land has been cleared and drained. The batter-board layout has been found to be satisfactory in the preliminary phases of construction. This method consists of placing batter-boards at each corner approximately 2 to 6 feet outside of the corner. Nails are set on the boards in such a way that when strings are extended between the nails, the strings outline the building area.

Stockpiling of Construction Material

Concrete Materials. In operations where large quantities of concrete are required, an aggregate batching plant and a cement batching plant are essential. Stockpiles of aggregates should be built up and maintained at the batching plant and at the crushing and screening plant. Stockpiles prevent shortages at the batching plant caused by temporary production or transportation difficulties and also provide an opportunity for the fine aggregates to reach a fairly stable and uniform moisture content and bulking factor. Large stockpiles are usually rectangular for ease in computing volumes and flat-topped to retain uniformity of gradation and to avoid the segregation which occurs when material is dumped so it can run down a long slope. An adequate supply of cement should be maintained at the cement batching plant. Consideration of the amount of concrete to be used in the project and the rate of placement will aid in determining the size of stockpile to be maintained. If admixtures are to be used, a sufficient quantity must be on hand.

Materials at Construction Site. An adequate supply of material for formwork construction should be stockpiled at the construction site. If scaffolding will be needed, it will be necessary to stockpile this material also. The size and quantity of lumber stored will depend on the type of forms and/or scaffolding to be used in the construction.

Batching Plants

The cement batching plant may be operated at the same location as the aggregate batching plant or it may be located closer to the mixer, depending on a number of conditions. After a layout has been developed, the batching plant is placed within crane reach of aggregate stockpiles and astride the batch truck routes. The crushing and screening plant is normally located at the pit but it may be operated at the batching plant or at a separate location. The initial location of aggregate, cement, and water, the quality of the aggregate, and the location of the work may all affect the position of the batching plant. If the road is favorable, a hillside location permits gravity-handling of materials without excessive new construction and may eliminate cranes or conveyors.

Safety Facilities

The planning of safety facilities and in some cases the actual construction of the facilities should take place during site preparation. Such facilities include overhead canopies and guard rails to protect personnel from falling debris and to prevent anyone from accidentally falling into open excavations. Certain sites, such as those where landslides may occur, will require additional safety facilities.

EXCAVATION

Once the construction site has been cleared and drained, the land must be cut to the proper elevation for the placement of footings. Initial excavation should be done with suitable equipment; however, final excavation should be done by hand, to the prescribed depths. Excavation should extend beyond the outside edge of the walls to allow for placement of forms and application of waterproofing. If too much material is excavated, place the concrete to the depth actually excavated. Do not refill excavations to the specified depth before placing the concrete because it is too difficult to compact the fill surface properly. Whenever an excavation is to be carried to a depth which may render the slopes unstable, some type of lateral support must be provided. This is true from the standpoint of safety as well as economy. Thus, it is good engineering practice to provide shoring whenever the stability of a slope is questionable. The type of shoring to be used will vary with the depth and size of the excavation, the physical characteristics of the soil, and the fluid pressure under saturated conditions. Sandy soils and wet earth generally require more extensive shoring than do firmer soils.

Hand Excavation

The rate of hand excavation is determined from table 5–1. It varies with the type of soil and the depth of excavation. When mechanical equipment is used, the last six inches of bottom excavation

Table 5–1. Earth Excavation by Hand

Type of material	Cubic yards per man-hour					
	Excavation with pick and shovel to depth indicated				Loosening earth—man with pick	Loading in trucks or wagons—one man with shovel and loose soil
	0 to 3 foot	0 to 6 foot	0 to 8 foot	0 to 10 foot		
Sand	2.0	1.8	1.4	1.3	—	1.8
Silty sand	1.9	1.6	1.3	1.2	6.0	2.4
Gravel, loose	1.5	1.3	1.1	1.0	—	1.7
Sandy silt-clay	1.2	1.2	1.0	.9	4.0	2.0
Light clay	.9	.7	.6	.7	1.9	1.7
Dry clay	.6	.6	.5	.5	1.4	1.7
Wet clay	.5	.4	.4	.4	1.2	1.2
Hardpan	.4	.4	.4	.3	1.4	1.7

*Table 5–2. Earth Excavation by Machine**

Equipment	Type of materials	Average output cu yd/hr
Power shovel (½ cu yd capacity)	Sandy loam	70
	Common earth	60
	Hard clay	45
	Wet clay	25
Short-boom dragline (½ cu yd capacity)	Sandy loam	65
	Common earth	50
	Hard clay	40
	Wet clay	20
Backhoe (⅓ cu yd capacity)	Sandy loam	55
	Common earth	45
	Hard clay	35
	Wet clay	25

*90° swing, closed pit such as basement.

must be cleared out and shaped by hand because it is extremely difficult to accurately excavate by machine.

Machine Excavation

Machine excavation is a necessity for large projects which require substantial excavation. Types of excavation equipment which are particularly suitable for use in concrete construction work include power shovels, dragline buckets, and backhoes. Considerations which enter into the selection of equipment are the total yardage to be moved, working time available, type of excavation, and nature of the area. Due to the many variables, it is not possible to give generalized rates of excavation for various types of equipment. Some typical rates of excavation for specific conditions are given in table 5–2. In practice there will be considerable variation from these rates.

FORMWORK

In connection with a job analysis, it is important to have a working knowledge of the equipment necessary for the job and an idea of how much work can be turned out by the form builders per unit of time.

Techniques. Standardized methods of making, erecting, and stripping forms should be developed to the maximum possible extent. This saves time and material and simplifies design problems.

Equipment. The required tools should be made readily available. An average job will require claw hammers, pinchbars, hand saws, portable electric saws, a table saw, levels, plumb lines and carpenter's squares.

Time Element

A carpenter of average skill can build and erect 10 square feet of wood forms per hour. This figure increases as the worker becomes more skilled in form construction. It also varies with the available tools and materials and the type of form. Some forms, such as stairways, require considerable physical support from below, and so require more manhours and materials than simpler forms. Additional time is required if it is necessary for the carpenters to move frequently from one level to another. An increase in manpower support at the ground level can result in increased efficiency.

MIXING

Established and well-defined concrete mixing procedures must be followed if the finished concrete is to be of good quality. Oversights in this phase of concrete construction whether through lack of competent and conscientious supervision or inattention to detail cannot be overcome later. It is the responsibility of those in charge of construction work to become familiar with concrete mixing procedures and to make sure that they are followed. The extra effort and care required are small in relation to the benefits.

Measuring Mix Materials

To produce concrete of uniform quality, the ingredients must be measured accurately for each batch. Materials should be measured within this percentage of accuracy: cement, 1 percent; aggregate, 2 percent; water, 1 percent; admixtures, 3 percent. Equipment should be capable of measuring quantities within these tolerances for the smallest batch regularly used as well as for larger batches. The accuracy of equipment should be checked periodically and adjusted when necessary. Admixture dispensers should be checked daily since errors in admixture measurements, particularly overdosages, can lead to serious problems in both fresh and hardened concrete.

Cement. Concrete mixes are normally designed using sacked cement as the unit. The use of bulk cement is common practice in commercial construction where large quantities are needed. It is stored in bins directly above a weighing hopper and discharged from the hopper. Since special equipment is required to transport bulk cement, sacked cement is used almost exclusively in troop construction, particularly in the theater of operations.

Aggregates. Aggregates for each batch should be measured accurately, either by weight or by volume. Measurement by weight is the most reliable method, since the accuracy of volume measurement depends on the accuracy of an estimate of the amount of bulking which varies according to the moisture in the sand. However, in some instances measurement by volume is more practical.

Measurement by weight. On comparatively small jobs the aggregate can be weighed on platform scales. The scales should be set on the ground, and runways constructed so that a wheelbarrow can be run onto one side of the scale and off the other, as shown in figure 5–1. With practice it is possible to fill a wheelbarrow so accurately that it is seldom necessary to add or remove material to obtain the correct weight. The amount of aggregate placed on each wheelbarrow should be the same and the quantity per batch should be supplied by an even number of wheelbarrow loads. Hence the wheelbarrow may not be loaded to capacity each time. Wheelbarrows of from 2- to 3-cubic foot capacity are available in engineer units.

Measurement by volume. Measuring by volume can be done by means of a 1-cubic foot measuring box built on the job. The inside of the box should be marked off in tenths of a cubic foot. The GI bucket can also be used as a measuring device; it contains 0.467 cubic feet which may be considered one-half cubic foot. If wheelbarrows are to be used to carry the aggregate from the storage pile to the mixer, the following procedure, based on a 3-cubic foot wheelbarrow, should be used. Assume that the proportions by volume are 1:2:3 and each batch is to contain three sacks of

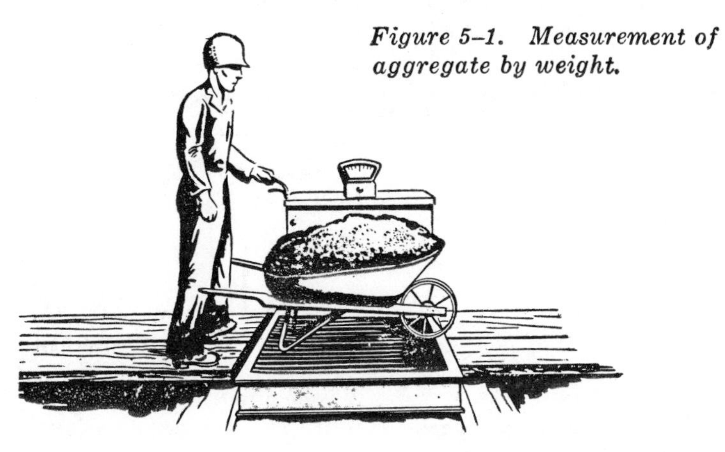

Figure 5–1. Measurement of aggregate by weight.

cement. Use the 1-cubic foot measuring box to load 2.0 cubic feet of sand in the wheelbarrow. Draw a line around the inside of the wheelbarrow at the level of the sand. Three wheelbarrows filled to this level will then be used per batch. If the sand is damp, bulking and an increase in volume will occur. For this reason measurement by volume must be undertaken with care. The coarse aggregate can be measured directly from the wheelbarrow, using the previously drawn line.

Water. Water for mixing must be accurately measured for every batch. If the aggregate contains excess moisture, this should be considered when mixing water is added. The water tanks of machine mixers are equipped with measuring devices operating on the principle of a siphon. After the tank is filled, the water is siphoned off. The

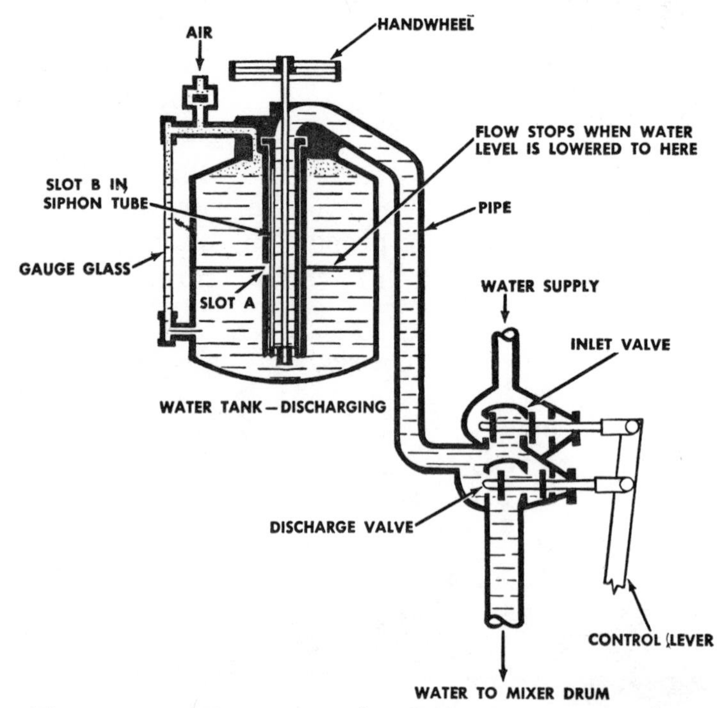

Figure 5–2. Water-measuring device on concrete mixer.

amount of water going to the mixer is determined by the setting of the siphon. Figure 5–2 illustrates this type of device. Slot B in the siphon tube is a vertical slot about one-half of an inch wide. Slot A is a spiral slot extending from the bottom to the top of the casing around the siphon tube. The handwheel turns the siphon tube which changes the point where the spiral slot crosses the vertical slot. One complete turn of the handwheel controls minimum to maximum quantity of water. When the water is lowered to the level indicated in the figure and where the slots cross, air enters the siphon tube, breaking the siphon and stopping the flow of water. Continual maintenance and periodic checks on the quantity of water delivered assure proper operation of the apparatus. If the mixer is not equipped with an automatic measuring device, a pail, marked for gallons and fractions, may be used to measure the water. In any event mixing water should be measured carefully.

Hand Mixing

Mixing is generally done by machine but some hand mixing is invariably necessary. A clean surface is required for this purpose. Ordinarily a wooden platform such as shown in figure 5–3, with close joints to prevent loss of mortar, is used. The wood surface should be moistened prior to mixing. The platform should be leveled before mixing is started. A clean, even paved surface will also serve the purpose of a mixing platform. The measured quantity of sand is placed on the bottom and the cement is spread over the sand and then the coarse aggregate is spread on top. Either a hoe or a square pointed D-handle shovel can be used to mix the materials. The dry materials should be turned at least three times until the color of the mixture is uniform. Water is added slowly while the mixture is turned again at least three times. Water is gradually added until the proper consistency is obtained. When two men are mixing they should face each other working their way through the pile and keeping the shovels close to the surface of the platform while turning the materials. One man can mix 1 cubic yard of concrete by hand in about an hour, but this is not an economical method of mixing concrete in batches of over 1 cubic yard. Concrete is also frequently mixed in a hoe-box as shown in figure 5–3.

Machine Mixing

Power concrete mixers are available in several sizes and types. A mixer will normally produce a batch about every 3 minutes, including charging and discharging. The actual hourly output of mixers may vary from 10 to 20 batches per hour.

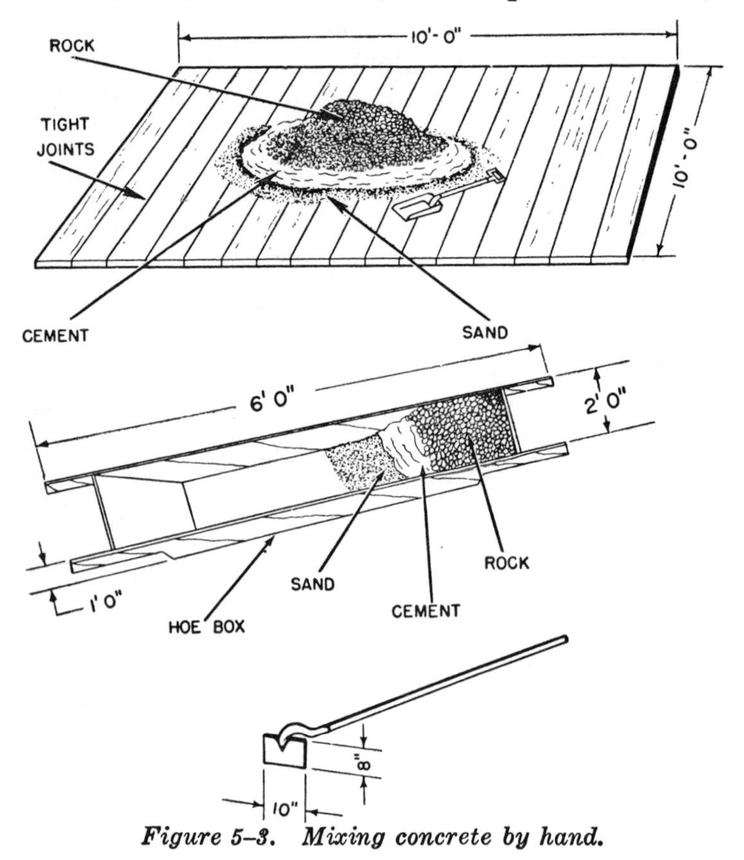

Figure 5–3. Mixing concrete by hand.

Types. The cubic-foot rating of a mixer is usually the number of cubic feet of wet concrete the machine mixes satisfactorily in one batch, except that most mixers will handle a 10 percent overload. The 16-S, power-driven skip mixer is TOE equipment in engineer construction battalions and is well suited for troop construction projects. The physical characteristics of this mixer are given in table 5–3.

Trailer mounted mixer. Trailer mounted mixers are commonly used for patches in the repair of concrete pavements and for fillets and curve widening in concrete pavement construction. A battery of trailer mounted mixers may serve as a central mix plant for large scale operations or they may be used in conjunction with a central mix plant.

Table 5–3. Physical Characteristics of the 16–S Mixer

Physical characteristics	Mixer, concrete, gas driven liquid cooled, end dis, charge, trailer mounted 4 pneumatic tired wheels, 16 cu ft
Drum capacity	16 cu ft
Hourly production	10 cu yd
Rating (sacks per batch)	2
Power unit:	
Horsepower	26
Fuel consumption (gph)	0.5
Water tank:	
Supply (gal)	None
Measuring (gal)	26
Drum dimensions:	
Diameter (in.)	57
Length (in.)	46
Overall dimensions (in.):	
Length	158
Width	96
Height	119
Weight (lb)	7,150

Truck mixer. Truck mixers are used to deliver concrete from a centrally located stationary mixer to the construction site and for picking up materials at a batching plant and mixing the concrete enroute to the job site. The truck mixer may pick up the concrete from the stationary mixer in a partially mixed state or in a completely mixed state in which case it functions as an agitator.

Stationary mixer. Stationary mixers include both on-site mixers and central mixers in ready mix plants. They are available in various sizes and may be of the tilting or nontilting type, or of the open-top revolving blade or paddle type.

Central Mix Plants. Stationary mixers or a battery of trailer mounted mixers are usually operated at a central mix plant. Central mix plants are normally set up for gravity-feed operation; a clamshell bucket crane, conveyor belts, or elevators carry materials to a batching plant set high enough to discharge directly into the loading gate or skip of the mixer, with the mixer discharging directly into dump trucks or other distributing equipment. Mixing time and mixing requirements do not differ appreciably from those previously discussed, but special control is required to insure that the concrete has the proper characteristics and workability upon arrival at the work site. Care must be taken to avoid segregation when using dump trucks.

Organization of Mixing Equipment and Materials. The proper placement of the mixing equipment and materials for the mix can result in large savings of time and labor. The mixer should be located as close as possible to the main section of the pour. On a concrete wall project, the mixer may be moved to each wall so as to reduce the distance the plastic concrete must be transported.

The gravel and sand should be stored as close as possible to the mixer without interfering with the transportation of the plastic concrete. The location of the water depends on the type of water supply. If the water is being piped in, a hose may be used to bring the water to a barrel near the mixer. If a water truck or trailer is used, it should be located next to the mixer. A typical arrangement of mixer and materials is shown in figure 5–4.

Operation of 16-S Mixer. Ten enlisted men and one noncommissioned officer are required for normal operation of the 16-S mixer. This crew can handle aggregate, sand, cement, and water, and operate the mixer. The noncommissioned officer must be competent to supervise the overall operation. This crew and equipment can produce about 10 cubic yards of concrete per hour, depending on the experience of the crew, the location of materials, and the rate of discharge of the mixer. At least a platoon of men would be required for an overall project similar to the one depicted in figure 5–4.

Charging the mixer. There are two ways of charging concrete mixers, by hand and with the mechanical skip. The 16-S mixer is equipped with a mechanical skip, as shown in figure 5–5. The cement, sand, and gravel are placed in the skip and then dumped into the mixer together while the water runs into the mixing drum on the side opposite the skip. The mixing water is measured from a storage tank on top of the mixer a few seconds before the skip is dumped to wash the mixer between batches. The coarse aggregate is placed in the skip first, the cement next and the sand is placed on top to prevent excessive loss of cement as the batch enters the mixer.

Discharging the mixer. When the material is ready for discharge from the mixer, the

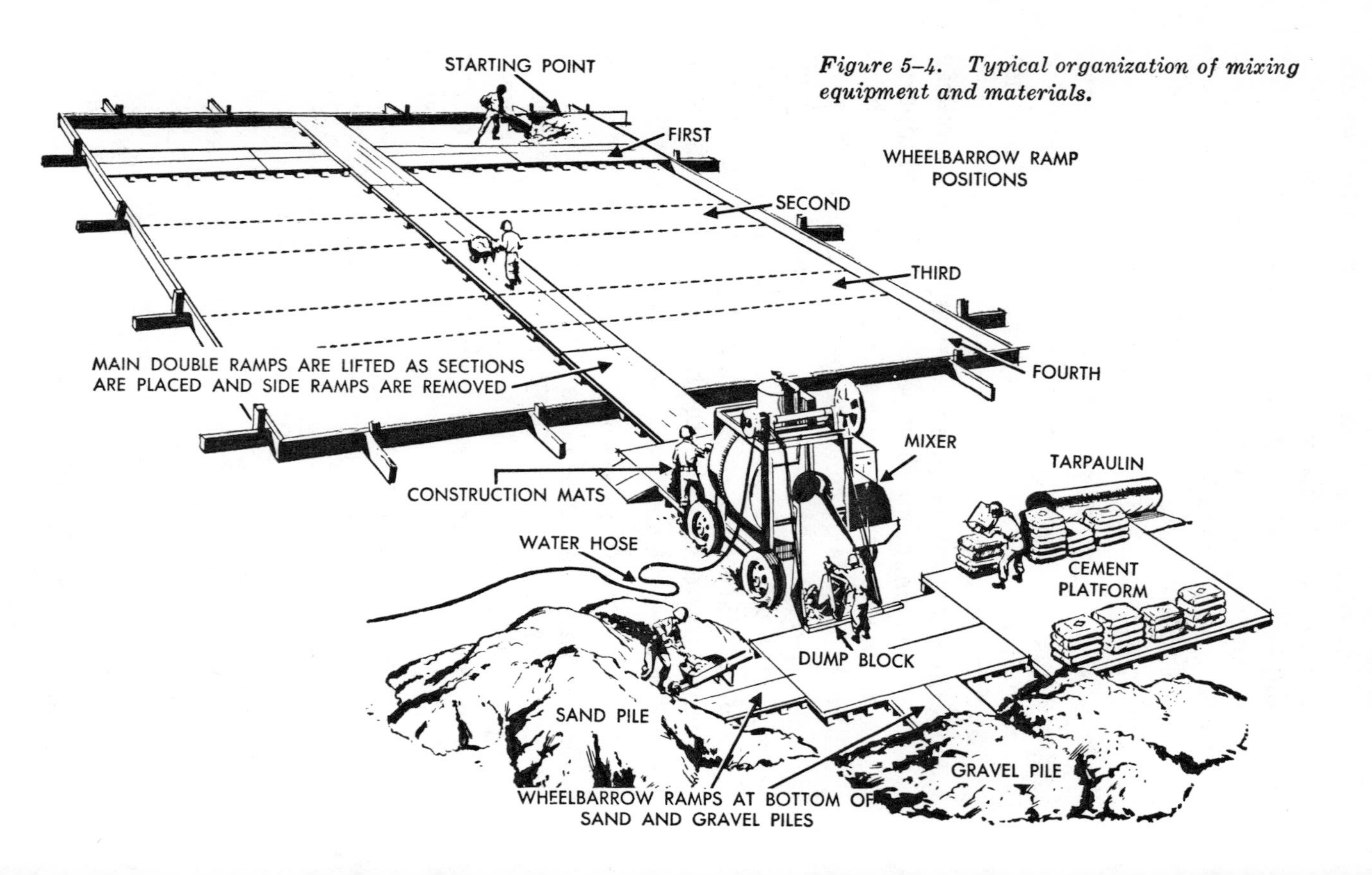

Figure 5–4. Typical organization of mixing equipment and materials.

discharge chute is moved into place to receive the concrete from the drum of the mixer. In some cases, dry concrete has a tendency to carry up to the top of the drum and not drop down in time to be deposited on the chute. Very wet concrete may not carry up high enough to be caught by the chute. This condition can be corrected by adjusting the speed of the mixer. For very wet concrete, the speed of the drum should be increased and for dry concrete, it should be slowed down.

Mixing time. The mixing time is determined from the time the water is added to the mixture. All mixing water should be added in the first quarter of the mixing period. The minimum mixing time per batch of concrete is 1 minute unless the batch exceeds 1 cubic yard. An additional 15 seconds of mixing time is required for each additional ½ cubic yard of concrete or fraction thereof.

Cleaning and maintaining the mixer. The mixer should be cleaned daily when it is in continuous operation or following each period of use if it is in operation less than a day. If the outside of the mixer is coated with form oil, the cleaning process can be speeded up. The outside of the mixer should be washed with a hose and all accu-

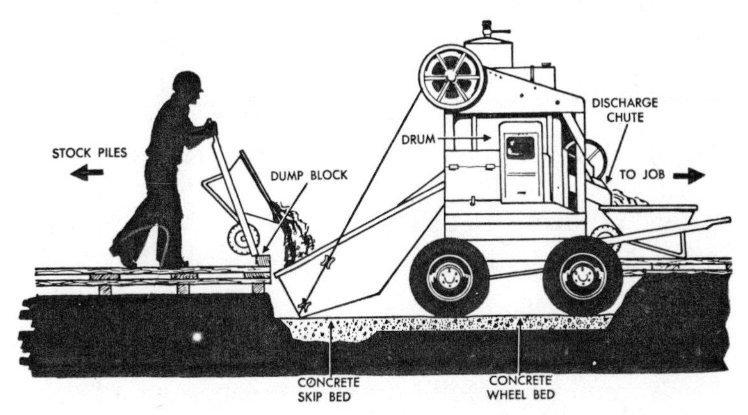

Figure 5–5. Charging a 16–S mixer.

mulated concrete should be knocked off. If the blades of the mixer become worn, or coated with hardened concrete, mixing action will be less efficient. Badly worn blades should be replaced. Hardened concrete should not be allowed to accumulate in the mixer drum. The mixer drum must be cleaned out whenever it is necessary to shut down for more than 1½ hours. Place a volume of coarse aggregate in the drum equal to one-half the capacity of the mixer and allow it to revolve for about 5 minutes. Discharge the aggregate and flush out the drum with water. Do not pound the discharge chute, drum shell, or skip to remove aggregate or hardened concrete, for concrete will adhere more readily to the dents and bumps created.

Remixing Concrete

Fresh concrete that is left standing tends to stiffen before the cement has hydrated to its initial set. Such concrete may be used if it becomes sufficiently plastic upon remixing that it can be compacted in the forms. Under careful supervision, a small increment of water may be added to these batches provided that all of the following conditions are met: maximum allowable water-cement ratio is not exceeded, maximum allowable slump is not exceeded, maximum allowable mixing and agitating time are not exceeded, and concrete is remixed for at least half the minimum required mixing time or number of revolutions. Indiscriminate addition of water is not acceptable since this lowers the quality of the concrete. Remixed concrete tends to harden rapidly. As a result, concrete placed adjacent to or above remixed concrete may cause a cold joint.

HANDLING AND TRANSPORTING

Each step in the handling and transporting of concrete must be carefully controlled to maintain uni formity within the batch and from batch to batch so that the completed work is consistent throughout. The method of handling and transporting concrete and the equipment used should not place restrictions on the consistency of the concrete. Consistency should be governed by the placing conditions. If these conditions allow a stiff mix, then equipment should be chosen which is capable of handling and transporting such a mix.

Requirements. The three main requirements for transportation of concrete from the mixing plant to the job sites are:

It must be rapid so that the concrete will not dry out or lose its workability or plasticity between mixing and placing.

Segreation of the aggregates and paste must be reduced to a minimum to assure uniform concrete. Loss of fine material, cement or water should be prevented.

Transportation should be organized so that there are no undue delays in the placing of concrete for any particular unit or section that would cause undesirable fill planes or construction joints.

Handling Techniques. Several general points concerning the handling of concrete are illustrated in fig. 5–6 to 5–8. Failure to observe the right procedures illustrated to prevent the segregation of the aggregates and paste can result in poor concrete in spite of good design and mixing procedures. Separation or segregation occurs because concrete is made up of materials of different particle size and specific gravity. The coarser particles in a concrete mix placed in a bucket tend to settle to the bottom and the water rises to the top. Honeycomb concrete or rock pockets are caused by the segregation of materials.

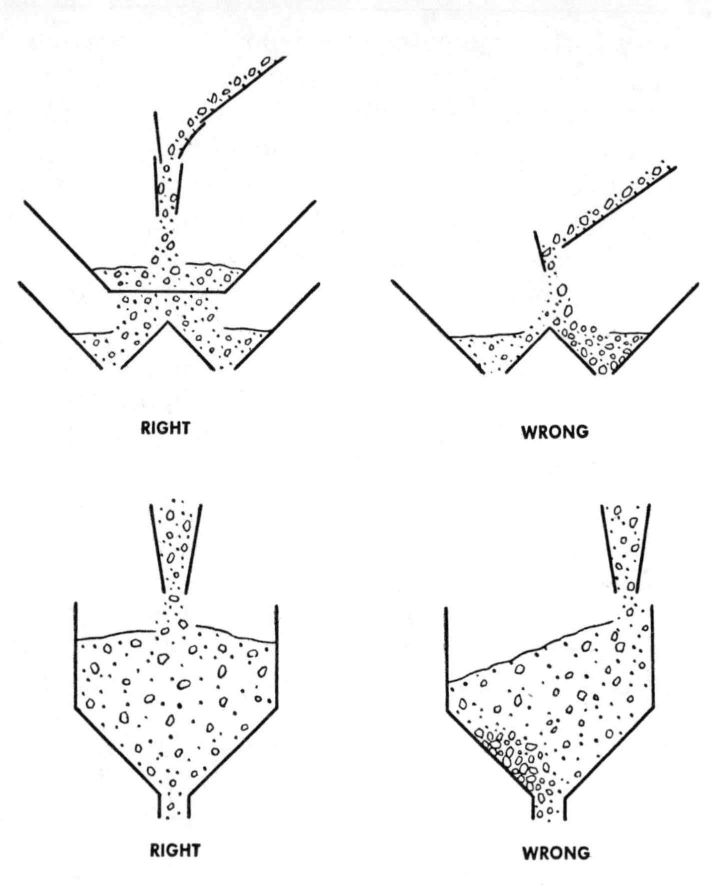

Figure 5–6. Concrete handling techniques.

Delivery Methods

Small Jobs. Equipment provided for the delivery of concrete should be based on 100-percent capacity of the mixer to take care of the peak draughts. Wheelbarrows or buggies are usually the most practical and economical means of delivering concrete for foundations, foundation walls, or slabs poured on or below grade. Power buggies (fig. 5–9), if available, are useful for longer runs. Hand buckets may be used to deliver small quantities of concrete as the occasion demands. When a situation arises requiring the con-

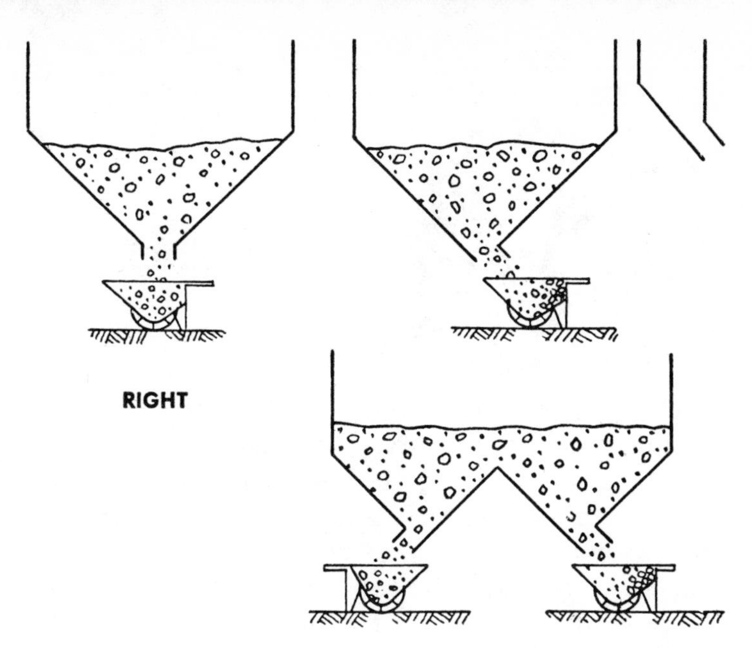

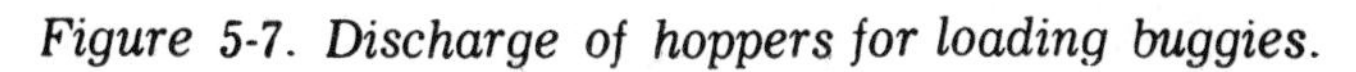

Figure 5-7. Discharge of hoppers for loading buggies.

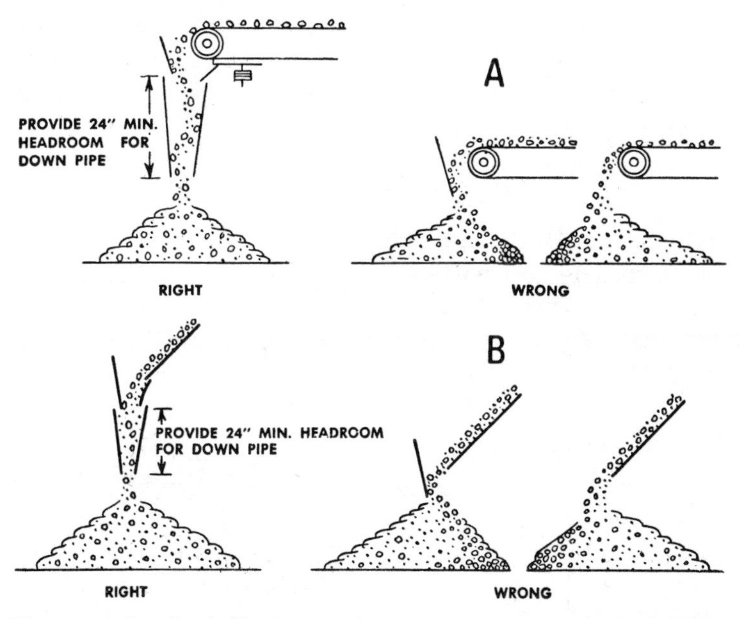

Figure 5-8. Control of separation at (A) end of conveyors and (B) end of chutes.

Figure 5–9. Power buggy.

crete to be poured approximately 15 feet above grade, inclined runways (fig. 5–10) can be constructed economically so that buggies or wheelbarrows can be used. When concrete is deposited below or at approximate grade, 2-inch plank runways placed on the ground permit the concrete to be poured directly into the form. If the difference in elevation from the runway to the bottom of the structure to be poured is large a chute similar to the illustration (fig. 5–11) should be used to avoid segregation. Usually the slope of the chute should be 2:1 or steeper for stiff mixes. If concrete is to be transported by buggy or wheelbarrow, suitable runways must be provided. A runway along a wall form is shown in figure 5–12. The top of the runway should be about level with the top of the form sheathing. In order to provide room for the ledger, the top wale should be kept at least 1 foot

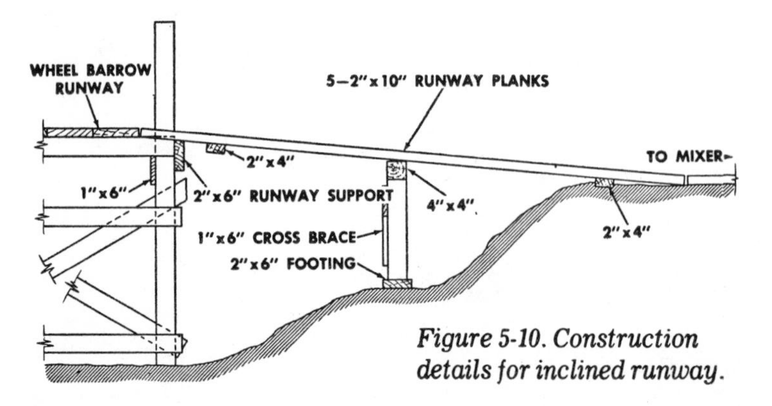

Figure 5-10. Construction details for inclined runway.

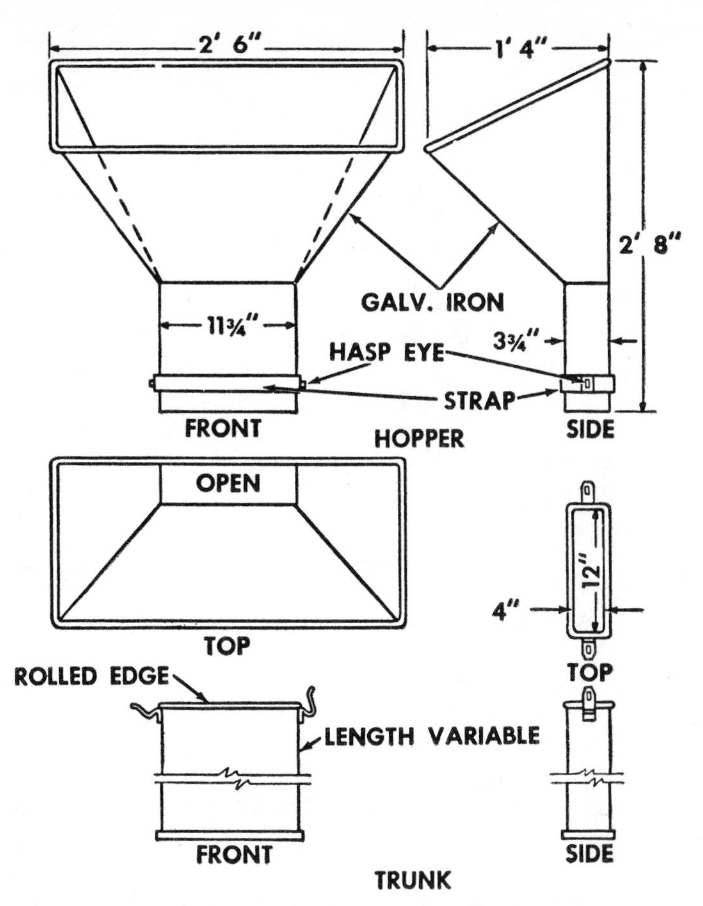

Figure 5-11. Chute to handle concrete.

below the top of the concrete. The runway should be made from rough lumber and should consist of 2- by 10-inch planks supported by 4 by 4's spaced on 6-foot centers. The 1- by 6-inch ledger on the form side should be nailed to the studs on the wall form. Runways must always be securely braced in order to prevent failure. Concrete can be elevated to about 5 to 6 feet above the level of the mixer by wheeling the concrete up inclined runways that have a slope of 10:1. If the concrete must be lifted more than 5 to 6 feet, and a large quantity is involved, it is more economical to use elevating

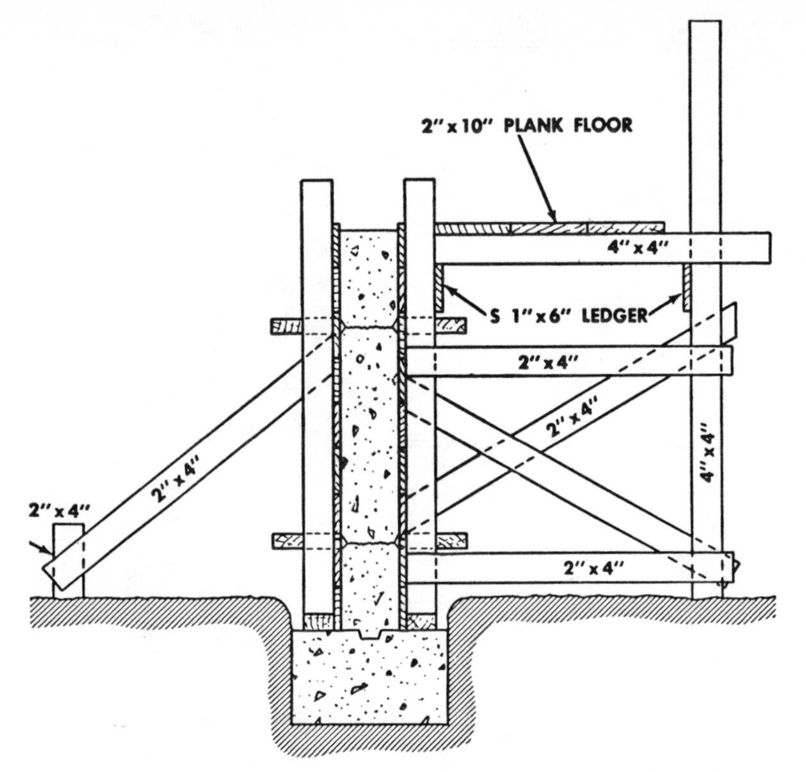

Figure 5-12. Runways for use of wheelbarrows or buggies.

equipment, such as a bucket and a crane. Runways for placing concrete on a floor slab should be supported from the form work or from the ground. These runways should also consist of 2 by 10's supported by 4 by 4's placed on 6-foot centers. If possible, all runways should be arranged so that the buggies or wheelbarrows will not have to pass each other on the runways at anytime.

Large Jobs. Transportation on large projects requires careful planning to assure adequate and timely delivery of concrete. Several types of equipment are used for this purpose.

Dump trucks. Dump trucks are commonly used to deliver concrete on large projects in the theater of operations. However, ordinary dump trucks are not designed as concrete carriers and care must be exercised in using them since prevention of segregation is not provided for as is the case with mixer or agitator trucks. Hauling distances are kept as short as possible. Segregation can be reduced by using a stiff mix and an air-entraining agent.

Agitator trucks. Transit-mix or ready-mix trucks can be used as agitator trucks to deliver premixed concrete. When used to deliver concrete only, as opposed to mixing enroute, the load capacity of the vehicle is 30 to 35 percent greater. However, the operating radius of the equipment when used as a delivery means for premixed concrete is somewhat more limited. The concrete can be discharged continuously or intermittently from a spout or chute which is moved from side to side.

Pumps. Pumps are useful when limited space prevents other more conventional means of delivery. This method involves the use of a heavy-duty piston pump to force concrete through a 6-, 7-, or 8-inch pipeline. Pumps operated by a 25 HP gasoline engine have a rated capacity of 15 to 20 cubic yards per hour. Larger equipment with a double acting pump has a capacity of 50 to 60 cubic yards per hour. These machines will force the above volume of concrete up to 800 feet horizontally or 100 feet vertically or any equivalent combination of these distances. A 90° bend in the pipeline corresponds to a loss of 40 feet horizontal delivery capacity and each foot of vertical lift corresponds to 8 feet of horizontal delivery distance. A mix with 2 or more inches of slump can be pumped. A light cement grout should be used in starting the pump to lubricate the pipeline. Maximum aggregate size for the 8-inch pipeline is 3 inches, for the 7-inch line, 2½ inches and for the 6-inch pipeline, 2 inches. The discharge line should be as straight as possible, with 5-foot radius bends. Steps should be taken to cool the line in hot weather so as to avoid impeded flow of concrete. An uninterrupted flow of fresh, plastic unseparated concrete of medium consistency must be delivered to the pump. Failure to keep an uninterrupted flow of concrete in the line can seriously delay the pour with resultant unwanted joints in the structure. A deflector or choke (restricted section) can be used at the discharge end of the line to direct and control the flow of concrete as required. The pump and line must be thoroughly flushed after each use.

PLACING

The full value of well-designed concrete cannot be obtained without proper placing and curing procedures. Good concrete placing and compacting techniques produce a tight bond between mortar and coarse aggregate and assure complete filling of the forms. These requirements are necessary if the full strength and best appearance of the finished concrete is to be realized.

Preliminary Preparation

General Preparation. Preparation prior to concreting includes compacting, trimming, and moistening the subgrade; erecting the forms; and setting the reinforcing steel. A moist subgrade is especially important in hot weather to prevent extraction of water from the concrete.

Rock Subgrades. When rock must be cut out, the surfaces in general should be vertical or horizontal rather than sloping. The rock surface should be roughened and thoroughly cleaned. Stiff

brooms, water jets, high-pressure air, or wet sandblasting may be used. All water depressions should be removed and the rock surface coated with a ¾-inch thick layer of mortar. The mortar should contain only fine aggregate and the water-cement ratio should be the same as for the concrete. It should have a 6-inch slump. Then work the mortar into the surface with stiff brushes.
rated without becoming muddy. The surface must be clean of debris and any dry loose material before concrete is placed.

Gravel and Sand Subgrades. Subgrades composed of gravel or other loose material need a tar paper or burlap cover before concrete is placed on them. Compacted sand need not be covered. However, it must be moist when the concrete is

Clay Subgrades. A subgrade composed of clay or other fine-grained soils should be moistened to a depth of 6 inches to aid in curing the concrete. If the soil is sprinkled intermittently, it can be satu-

placed to prevent the absorption of water from the concrete. Tar paper should be lapped not less than 1 inch, and stapled. Burlap should be joined by sewing with wire and moistened by sprinkling before concrete is placed on it.

Preparation of Forms. Shortly before concrete is placed, the forms should be checked for tightness and cleanliness. Bracing should be checked to assure that there will be no movement of the forms during placing. The forms are coated with a suitable form oil or coating material that will keep the concrete from sticking. In an emergency, the forms can be moistened with water to prevent sticking. Forms that have been exposed to the sun for some time will dry out and the joints will tend to open up. Saturating the forms with water helps to close such joints.

Depositing Fresh Concrete on Hardened Concrete. When new concrete is deposited on hardened concrete, the hardened concrete must be nearly level, clean, and moist, with some aggregate particles partially exposed, to obtain a good bond and a watertight joint. If there is a soft layer of mortar or laitance on the surface of the hardened concrete it should be removed. Wet sand blasting and washing is the most effective means of preparing the old surface where the sand deposit can be easily removed. Always moisten hardened concrete before placing new concrete. Dried-out concrete must be saturated for several hours. In no case should there be pools of water on the old surface when fresh concrete is deposited.

Depositing Concrete

Basic Considerations. Concrete should be deposited in even horizontal layers and should not be puddled or vibrated into place. The layers should be from 6 to 24 inches in depth depending on the type of construction. The initial set should not take place before the next layer is added. To prevent honeycombing or avoid spaces in the concrete, the concrete should be vibrated or spaded. This is particularly desirable in wall forms with considerable reinforcing. Care should be taken not to overvibrate, because segregation and a weak surface may result.

Maximum drop. There is a temptation, in the interests of time and effort, to drop the concrete from the point to which it has been transported regardless of the height of the forms, but the free fall of concrete into the forms should be reduced to a maximum of 3 to 5 feet unless vertical pipes, suitable drop chutes, or baffles are provided. Figure 5–6 suggests several methods to control the fall of concrete and prevent honeycombing of concrete, and other undesirable results.

Thickness of layers. Concrete should be deposited in horizontal layers whenever possible and each layer consolidated before the succeeding layer is placed. Each layer should be placed in one operation. In mass concrete work, where concrete is deposited from buckets, the layers should be from 15 to 20 inches thick. For reinforced concrete members the layers should be from 6 to 20 inches thick. The thickness of the layers depends on the width between forms and the amount of reinforcement.

Positioning. Concrete should be placed as nearly as possible in its final position. Horizontal movement should be avoided since this results in segregation because mortar tends to flow ahead of coarser material. Concrete should be worked thoroughly around the reinforcement and bedded fixtures, into the corners, and on the sides of the forms.

Rate of placing. On large pours so as to avoid excess pressure on forms, the rate of filling should not exceed 4 feet per hour measured vertically, except for columns. Placing will be coordinated so that the concrete is not deposited faster than it can be properly compacted. In order to avoid cracking during settlement an interval of at least 4 hours, preferably 24 hours, should elapse between completion of columns and walls and the placing of slabs, beams, or girders supported by them.

Wall Construction. For walls, the first batches should be placed at the ends of the section. Placing should then proceed toward the center for each layer, if more than one layer is necessary, to prevent water from collecting at the ends and corners of the forms. This method should also be used in placing concrete for beams and girders. For wall construction, the inside form should be stopped off at the level of the construction. Overfill the form for about 2 inches and remove the excess just before setting occurs to insure a rough, clean surface. Before the next lift of concrete is placed on this surface, a ½ to 1-inch thick layer of sand-cement mortar should be deposited on it. The mortar should have the same water content as the concrete and should have a slump of about 6 inches to prevent stone pockets and help produce a water tight joint. Proper procedure for placing concrete in walls is shown in figure 5–13. Note the use of drop chutes and port openings for placing concrete in the lower portion of the wall. The port openings are located at about 10-foot intervals along the wall. Concrete for the top portion of the wall can be placed as shown in A, figure 5–14 When pouring walls, remove the form spreaders as the forms are filled.

Slab Construction. For slabs, the concrete should be placed at the far end of the slab, each batch dumped against previously placed concrete, as shown in *B*, figure 5–14. The concrete should not be dumped in separate piles and the piles then leveled and worked together. Nor should the concrete be deposited in big piles and then moved horizontally to its final position, since this practice results in segregation.

Placing Concrete on Slopes. Procedure for placing concrete on slopes is shown in C, figure 5–14. Always deposit the concrete at the bottom of the slope first, and proceed up the slope as each batch is dumped against the previous one. Compaction is thus increased by the weight of the newly added concrete when it is consolidated.

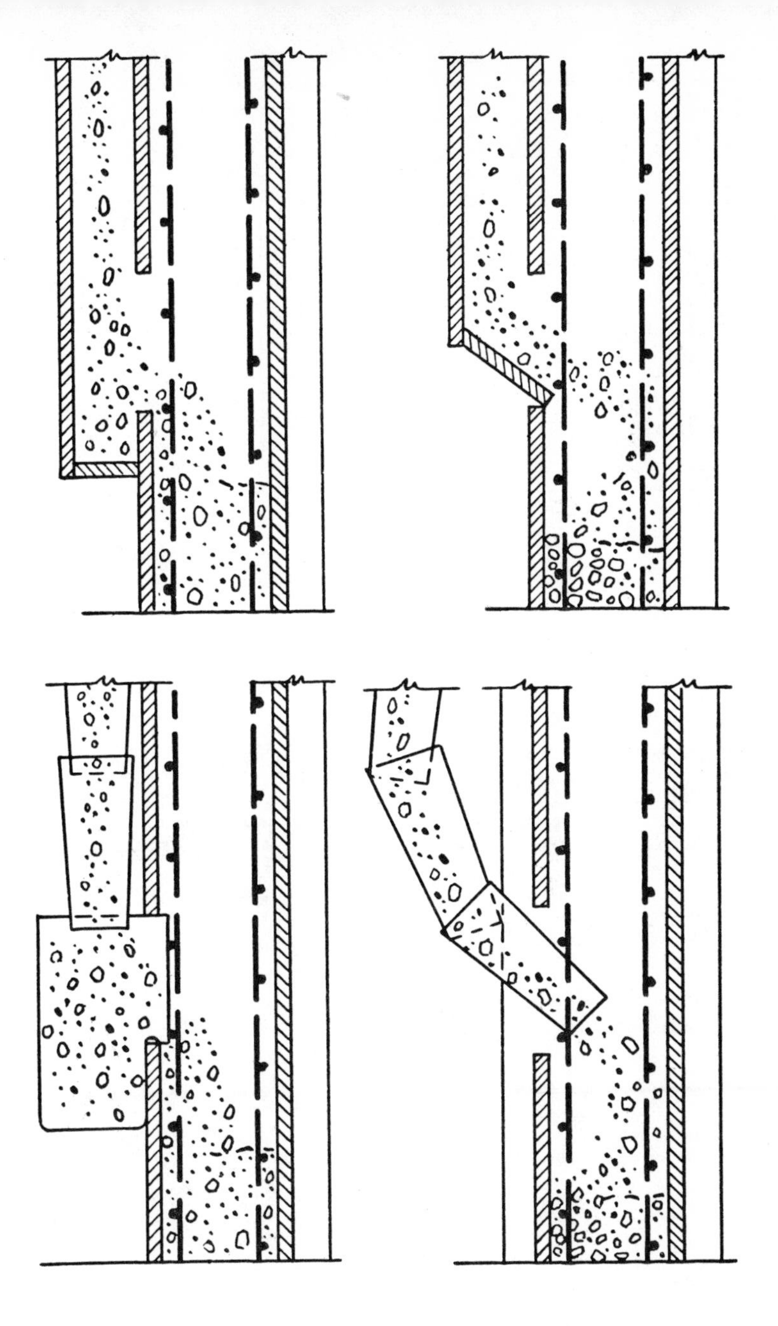

Figure 5–13. Placing concrete in high wall form.

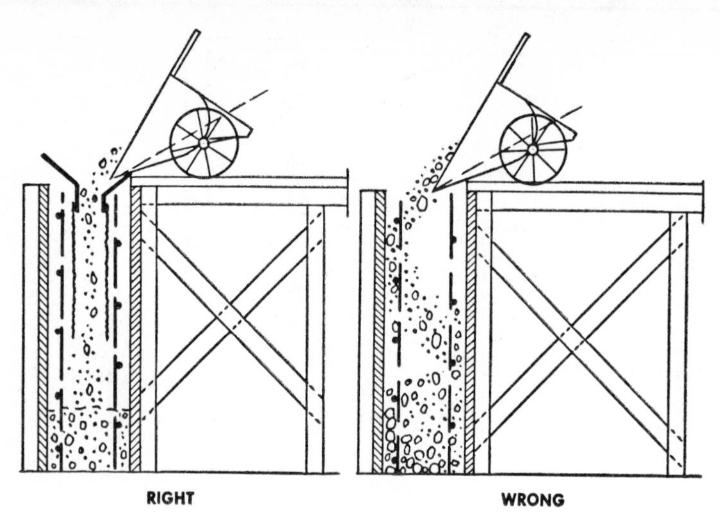

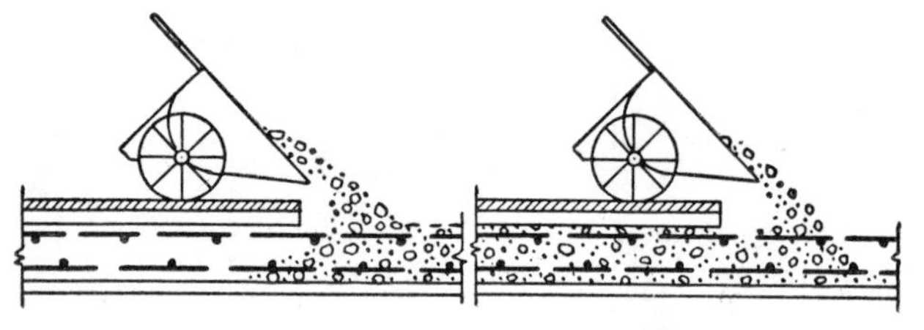

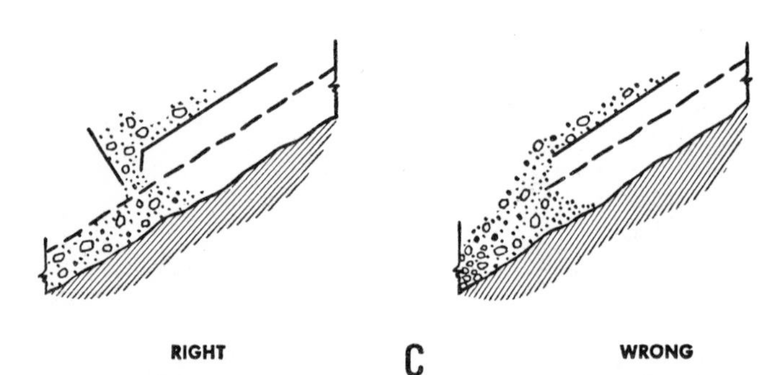

Figure 5-14. Placing concrete in (A) top of wall, B) slab, and (C) sloping surfaces.

Consolidating Concrete

Purpose. With the exception of concrete placed under water, concrete is compacted or consolidated after placing. Consolidation may be accomplished by the use of hand tools such as spades, puddling sticks, and tampers; but the use of mechanical vibrators is preferred. Compacting devices must reach the bottom of the form and must be small enough to pass between reinforcing bars. Consolidation eliminates rock pockets and air bubbles and brings enough fine material to the surface and against forms to produce the desired finish. In the process of consolidation the concrete is carefully worked around all reinforcing steel to assure proper embedding of the steel in the concrete. Displacement of reinforcing steel must be avoided since the strength of the concrete member depends on proper location of the reinforcement.

Vibration. Vibrators consolidate concrete by pushing the coarse aggregate down and away from the point of vibration. With vibrators it is possible to place concrete mixtures too stiff to be placed in any other way. In most structures, concrete with a 1- or 2-inch slump can be deposited. Stiff mixtures require less cement and are therefore more economical. Moreover, there is less danger of segregation and excessive bleeding. The mix must not be so stiff that an excessive amount of labor is required to place it. Internal vibrators are available in engineer construction battalions. The internal vibrator involves insertion of a vibrating element into the concrete. The external type is applied to the forms. It is powered by electric motor, gasoline engine, or compressed air. The internal vibrator should be inserted in the concrete at intervals of approximately 18 inches for 5 to 15 seconds to allow some overlap of the area vibrated at each insertion. Whenever possible the vibrator should be lowered vertically into the concrete and allowed to descend by gravity. The vibrator should pass through the layer being placed and penetrate the layer below for several inches to insure a good bond between the layers. Under normal conditions there is little likelihood of damage from the vibration of lower layers provided the disturbed concrete in these lower layers becomes plastic under the vibratory action. Sufficient vibration has taken place when a thin line of mortar appears along the form near the vibrator, when the coarse aggregate has sunk into the concrete, or when the paste just appears near the vibrator head. The vibrator should then be withdrawn vertically at about the same rate that it descended. The length of time that a vibrator should be left in the concrete depends on the slump of the concrete. Mixes that can be easily consolidated by spading should not be vibrated because segregation may occur. Concrete that has a slump of 5 or 6 inches should not be vibrated. Vibrators should not be used to move concrete any distance in the form. Some hand spading or puddling should accompany vibration.

Hand Methods. Hand methods for consolidating concrete include the use of spades or puddling sticks and various types of tampers. For consolidation by spading, the spade should be shoved down along the inside surface of the forms through the layer deposited and down into the lower layer for a distance of several inches, as shown in figure 5–15. Spading or puddling should continue until the coarse aggregate has disappeared into the mortar.

Placing Concrete Under Water

Suitable Conditions. Concrete should be placed in air rather than under water whenever possible. When it must be placed under water, the work should be done under experienced supervision and certain precautions should be taken. For best results, concrete should not be placed in water having a temperature below 45° F. and should not be placed in water flowing with a velocity greater than 10 feet per minute, although sacked concrete may be used for water velocities greater than this. If the water temperature is below 45° F., the temperature of the concrete when it is deposited should be above 60° F. but in no case above 80° F. If the water temperature is above 45° F., no temperature precautions need be taken. Coffer dams or forms must be tight enough to reduce the current to less than 10 feet per minute through the space to be concreted. Pumping of water should not be permitted while concrete is being placed or for 24 hours thereafter.

Tremie Method. Concrete can be placed under water by several methods, the most common of which is with a tremie. The tremie method involves a device shown in figure 5–16. A tremie is a pipe having a funnel-shaped upper end into which the concrete is fed. The pipe must be long enough to reach from a working platform above water level to the lowest point at which the concrete is to be deposited. Frequently the lower end of the pipe is equipped with a gate, permitting filling before insertion in water. This gate can be opened from above at the proper time. The bottom or discharge end is kept continuously buried in newly placed concrete, and air and water are excluded from the pipe by keeping it constantly filled with concrete. The tremie should be lifted slowly to permit the concrete to flow out. Care must be taken not to lose the seal at the bottom. If lost, it is necessary to raise the tremie, plug the lower end, and lower the

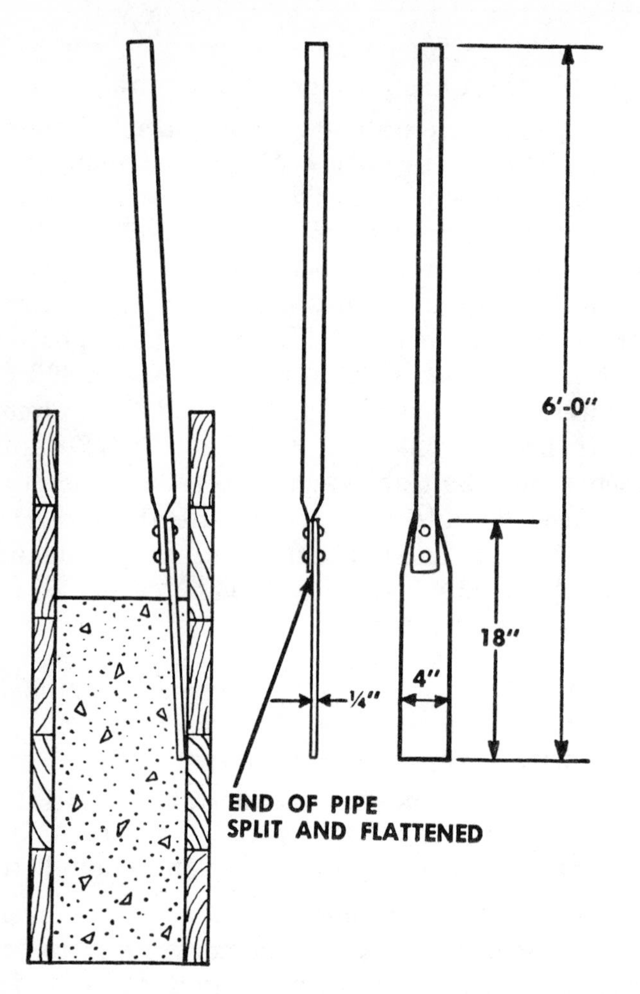

Figure 5–15. Consolidation by spading and the spading tool.

tremie into position again. The tremie should not be moved laterally through the deposited concrete. When it is necessary to move the tremie, it should be lifted out of the concrete and moved to the new position, keeping the top surface of the concrete as level as possible. A number of tremies should be used if the concrete is to be deposited over a large area. They should be spaced on 20- to 25-foot centers. Concrete should be supplied at a uniform rate to all tremies with no interruptions at any of them. Pumping from the mixer is the best method of supplying the concrete. Large tremies can be suspended from a crane boom and can be easily raised and lowered with the boom. Concrete that is placed with a tremie should have a slump of about 6 inches and a cement content of seven sacks per cubic yard of concrete. About 50 percent of the total aggregate should be sand and the maximum coarse aggregate size should be from 1½ to 2 inches.

Concrete Buckets. Concrete can be placed at considerable depth below the water surface by means of open-top buckets with a drop bottom. Concrete placed by bucket can be slightly stiffer than that placed by tremie but it should still con-

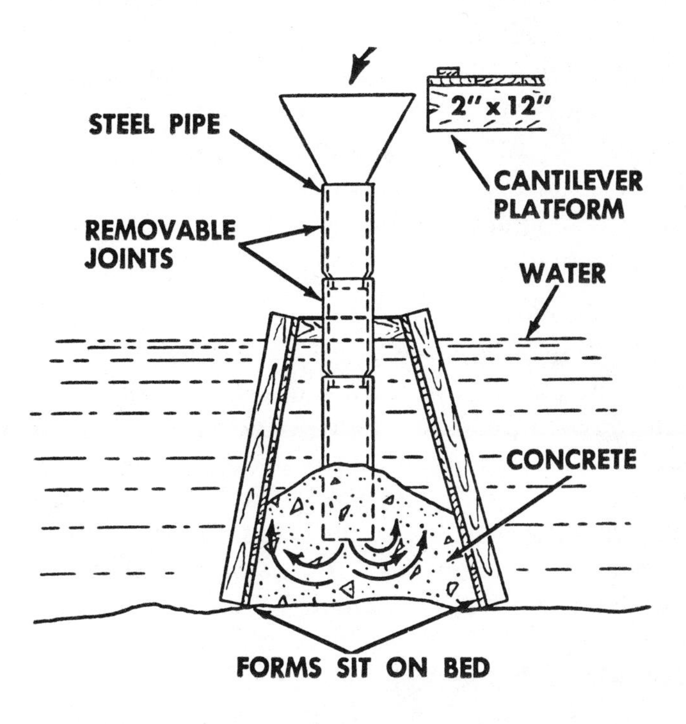

Figure 5–16. Placing concrete under water with a tremie.

tain seven sacks of cement per cubic yard. The bucket is completely filled and the top covered with a canvas flap. The flap is attached to one side of the bucket only. The bucket is lowered slowly into the water so that the canvas will not be displaced. Concrete must not be discharged from the bucket before the surface upon which the concrete is to be placed has been reached. Soundings should be made frequently so that the top surface is kept level.

Sacked Concrete. In an emergency, concrete can be placed under water in sacks. Jute sacks of about 1-cubic foot capacity, filled about two-thirds full, are lowered into the water, preferably shallow water. These sacks are placed in header and stretcher courses, interlocking the entire mass. A header course is placed so that the length of the sack is at right angles to the direction in which the stretcher-course sacks are laid. Cement from one sack seeps into adjacent sacks and they are thus bonded together. Experience has shown that the less the concrete under water is disturbed after placement, the better it will be. For this reason, compaction should not be attempted.

FINISHING

The finishing process provides the desired surface effect of the concrete. The concrete finishing process may be performed in many ways, depending on the effect required. Occasionally only correction of surface defects, filling of bolt holes or cleaning is necessary. Unformed surfaces may require only screeding to proper contour and elevation, or a broomed, floated, or troweled finish may be specified.

Finishing Operations

Screeding. After a floor slab, sidewalk or pavement has been placed, the top surface is rarely at the exact elevation desired. The process of striking off the excess concrete in order to bring the surface to the right elevation is called screeding. This operation can begin as soon as the concrete has been placed. Prior to screeding the concrete should be vibrated to lower larger sized aggregate to avoid interference with the screed. A templet with a straight lower edge if a flat surface is required, or curved if a curved surface is required, is moved back and forth across the concrete with a sawing motion. The templet rides on wood or metal strips that have been established as guides. With each sawing motion the templet is moved a short distance along the forms as shown in figure 5-17. There should be a surplus of concrete against the front face of the templet which will be forced into the low spots as the templet is moved forward. If there is a tendency for the templet to tear the surface, the rate of forward movement of the templet should be reduced or the bottom edge should be covered with metal. In most cases this will stop the tearing action. Such procedures are necessary when air-entrained concrete is used because of the sticky nature of this type of concrete. It is possible to hand screed surfaces up to 30 feet in width but for efficient screeding it is best not to go beyond 10 feet. Three men, excluding a vibrator operator, can screed approximately 200 square feet of concrete per hour. Two men operate the screed and the third man pulls excess concrete from the front of the screed. It is necessary to screed the surface twice to remove the surge of excess concrete caused by the first screeding.

Floating. If a smoother surface is required than the one obtained by screeding, the surface should be worked sparingly with a wood or metal float or finishing machine. A wood float is shown

Figure 5–17. Screeding operations.

in 1, figure 5–18 and the float in use is shown in 2, figure 5–18. This process should take place shortly after screeding and while the concrete is still plastic and workable. Floating should not begin until the water sheen has disappeared and the concrete has hardened sufficiently that a man's foot leaves only a slight imprint. The purpose of floating is threefold: to embed aggregate particles just beneath the surface; to remove slight imperfections, high spots, and low spots; and to compact the concrete at the surface in preparation for other finishing operations. The concrete must not be overworked while it is still plastic, to avoid bringing an excess of water and mortar to the surface. This fine material will form a thin weak layer that will scale or wear off under usage. Where a coarse texture is desired as the final finish, it is usually necessary to float the surface a second time after it has partially hardened so that the required surface will be obtained. In slab construction long-handled wood floats are used as shown in 3, figure 5–18. The steel float is used the same way as the wood float but it gives the finished concrete a much smoother surface. Steel floating should begin when the water sheen disappears from the concrete surface, to avoid cracking and dusting of the finished concrete. Cement or water should not be used to aid in finishing the surface.

Troweling. If a dense, smoother finish is desired, floating must be followed by steel troweling at some time after the moisture film or sheen disappears from the floated surface and when the concrete has hardened enough to prevent fine material and water from being worked to the surface. This step should be delayed as long as possible. Excessive troweling too early tends to produce crazing and lack of durability; too long a delay in troweling results in a surface too hard to finish properly. The usual tendency is to start to trowel too soon. Troweling should leave the surface smooth, even, and free of marks and ripples. Spreading dry cement on a wet surface to take up excess water is not good practice where a wear-resistant and durable surface is required. Wet spots must be avoided if possible; when they do occur, finishing operations should not be resumed until the water has been absorbed, has evaporated, or has been mopped up. A surface that is fine-textured but not slippery may be obtained by troweling lightly over the surface with a circular motion immediately after the first regular troweling. In this process, the trowel is kept flat on the surface of the concrete. Where a "hard steel-troweled finish" is required, the first regular troweling is followed by a second troweling after the concrete has become hard enough so that no mortar adheres to the trowel and a ringing sound is produced as the trowel passes over the surface. During this final troweling, the trowel should be tilted slightly and heavy pressure exerted to thoroughly compact the surface. Hair cracks are usually due to a concentration of water and fines at the sur-

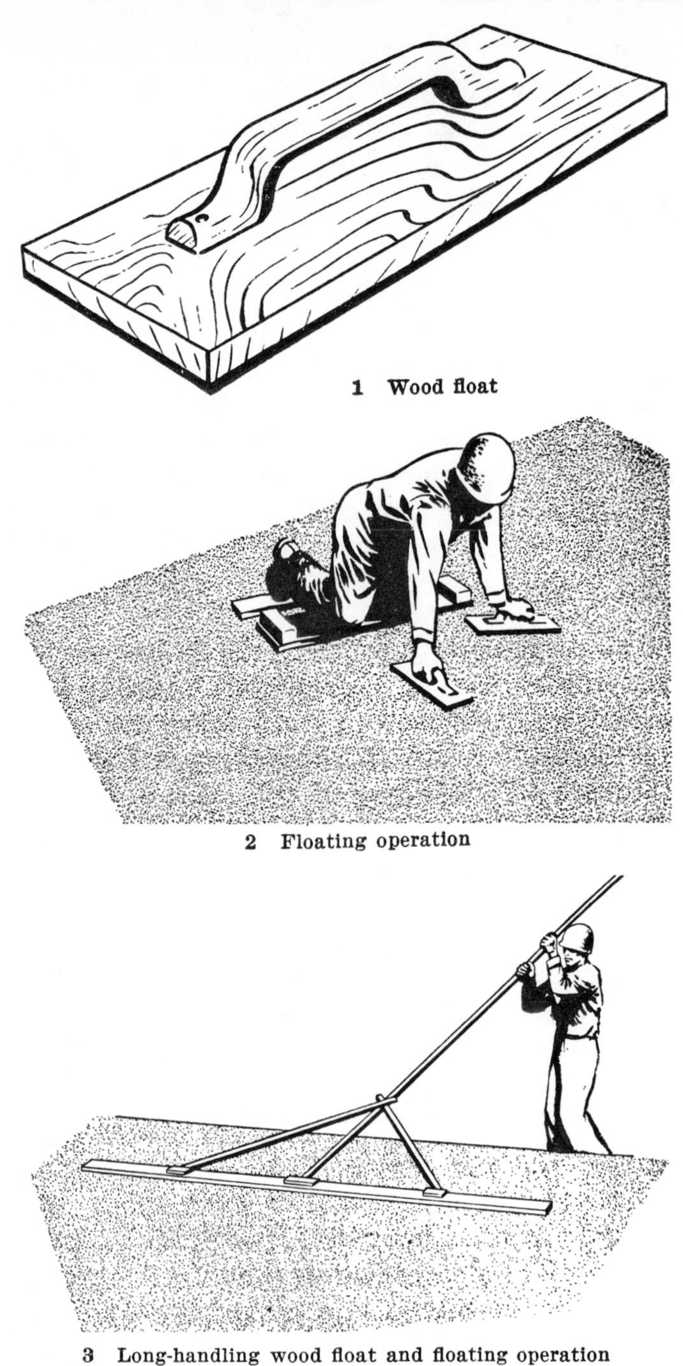

1 Wood float

2 Floating operation

3 Long-handling wood float and floating operation

Figure 5–18. Wood floats and floating operations.

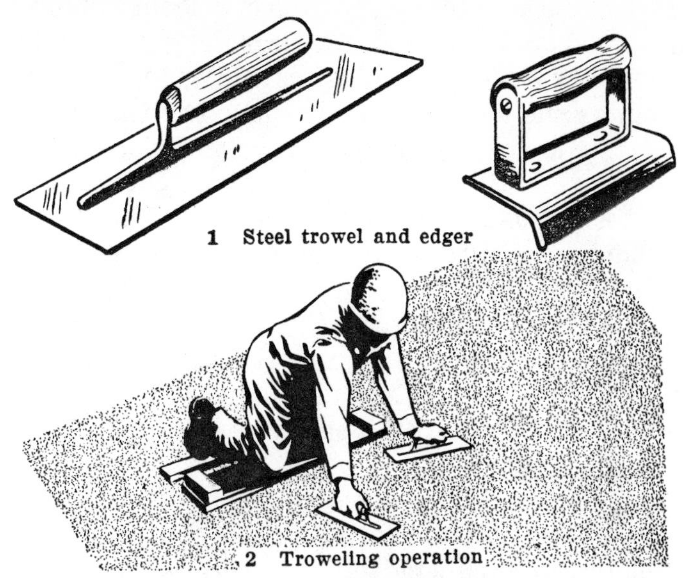

Figure 5–19. Steel finishing tools and troweling operations

face resulting from overworking the concrete during finishing operations. Such cracking is aggravated by too rapid drying or cooling. Cracks that develop before troweling usually can be closed by pounding the concrete with a hand float. A steel trowel and an edger are shown in 1, figure 5–19 and the troweling operation in 2, figure 5–19.

Brooming. A nonskid surface can be produced by brooming the concrete before it has thoroughly hardened. Brooming is carried out after the floating operation. For some floors and sidewalks where severe scoring is not desirable, the broomed finish can be produced with a hair brush after the surface has been troweled to a smooth finish once. Where rough scoring is required, a stiff broom made of steel wire or coarse fiber should be used. Brooming should be done in such a way that the direction of the scoring is at right angles to the direction of the traffic.

Rubbed Finish. A rubbed finish is required when a uniform and attractive surface must be obtained although it is possible to produce a surface of satisfactory appearance without rubbing if plywood or lined forms are used. The first rubbing should be done with coarse carborundum stones as soon as the concrete has hardened so that the aggregate is not pulled out. The concrete should then be cured until final rubbing. Finer carborundum stones are used for the final rubbing. The concrete should be kept damp while being rubbed. Any mortar used to aid in this process and left on the surface should be kept damp for 1 to 2 days after it sets in order to cure properly. The mortar layer should be kept to the minimum as it is likely to scale off and mar the appearance of the surface.

Finishing Pavement

Machine Finishing. Machine finishing is carried out at such time as the concrete takes its initial set. The concrete must, however, be in workable condition at the time of the finishing operation. The screeds and vibrator on the machine finisher are set to give the proper surface elevation and produce a dense concrete. In most cases, there should be a sufficiently thick layer of mortar ahead of the screed to insure that all low spots will be filled. The vibrator follows the front screed and the rear screed is last. The rear screed should be adjusted to carry enough grout ahead of it to insure continuous contact between screed and pavement. If forms have been set in good alinement and firmly supported, and if the concrete has the right workability, no more than two passes of the machine should be required to produce a satisfactory surface. TM 5–331 describes operation and maintenance of machine finishers.

Hand Finishing. Hand finishing with floats behind the finishing machine is often more harmful than beneficial. Excessive manipulation of the concrete is harmful to the wearing surface. Scaling has often been attributed to this. It is sometimes necessary to use a longitudinal float to decrease longitudinal variations in the surface. Such a float is made of wood, 6 by 10 inches wide and 12 to 18 feet long, fitted with a handle at each end and operated by two men on form-riding bridges. The float is oscillated longitudinally as it is moved transversely. A 10-foot straightedge pulled from the center of the pavement to the form will remove any minor surface irregularities and laitance. Unless considerable care is exercised as the straightedge or float approaches the form, it will ride up on the concrete resulting in a hump in the surface, especially where construction and expansion joints occur. The surface should have no coating of weak mortar or scum that will later scale off. After the water sheen disappears, the final surface finish is applied by dragging a clean piece of burlap longitudinally along the pavement strip. This is known as belting and is done by two men, one on each side of the forms. A non-skid surface is obtained by stroking with bassine brooms having fibers about 4½ inches long. The grooves cut by the broom should not be over 3/16-inch deep. All corners of the paving should be rounded with an edging tool. Expansion joints must be cleaned out and prepared for filling.

Repairing Concrete

After forms are removed, small projections must be removed, tierod holes filled and honeycombed areas repaired. These repairs should be made as soon as possible after the forms are removed. The repair of concrete is covered in detail in section XII of this chapter.

Cleaning Concrete

Concrete surfaces are not always uniform in color when forms are removed. If appearance is important the surface should be cleaned.

Cleaning With Mortar.

The surface can be cleaned with a cement-sand mortar consisting of one part portland cement and one and one-half to two parts fine sand. The mortar should be applied to the surface by a brush after all defects have been repaired. If a light-colored surface is desired, white portland cement can be used. The surface should be scoured vigorously with a wood or cork float immediately after applying the mortar. Excess mortar should be removed with a trowel after 1 or 2 hours which gives time for the mortar to harden enough so the trowel will not remove it from the small holes. After the surface has dried, it should be rubbed with dry burlap to remove any loose material. No visible film of mortar should remain after the rubbing. One section should be completed without stopping. Mortar left on the surface overnight is very difficult to remove.

An alternate method of cleaning with mortar consists of rubbing the mortar over the surface with clean burlap. The mortar should have the consistency of thick cream and the surface should be almost dry. The excess mortar is then removed by rubbing with clean burlap. Removal should be delayed long enough to prevent smearing but should be completed before the mortar hardens. The mortar is allowed to set several hours, then cured for 2 days. After curing, the surface is permitted to dry and is vigorously sanded with No. 2 sandpaper. This removes all excess mortar not removed by the sack rubbing and produces a surface of uniform appearance. For best results, mortar cleaning should be done in the shade on a cool, damp day.

Sandblasting. Surface stains, particularly rust, can be completely removed by lightly sandblasting the surface. This method is more effective than washing with acid.

Acid Cleaning. Acid washing can be used where the staining is not severe. Acid washing should be preceded by a two-week period of moist curing. The surface is first wetted and, while still damp, is scrubbed thoroughly with a 5 to 10 percent solution of muriatic acid using a stiff bristle brush. The acid is removed by immediate, thorough flushing with clean water. If possible, acid washing should be followed by four additional days of moist curing. When handling the acid, goggles are worn to protect the eyes and precautions must be taken to prevent the acid from contacting hands, arms, and clothing.

CURING

Hydration. The addition of water to portland cement and the formation of a water-cement paste starts a chemical reaction whereby the cement becomes a bonding agent. During this process, known as hydration, the main compounds of portland cement and water react to form products of hydration which produce a firm and hard substance—the hardened cement paste. The rate and degree of hydration, and as a result the strength of the concrete, are dependent on the curing process followed after the concrete has been placed and consolidated. The process of hydration continues for an indefinite period at a decreasing rate as long as water is in the mixture and temperature conditions are favorable. Once the water is removed, hydration ceases and cannot be restarted.

Importance. Curing refers to the steps necessary to keep concrete moist and as near to 73° F. as practicable until it has reached its design strength. Properties of concrete such as resistance to freezing and thawing, strength, watertightness, wear resistance, and volume stability improve with age as long as these conditions, which are favorable for continued hydration, are maintained. It follows that concrete should be protected so that moisture is not lost during the early hardening period and that the concrete temperature is kept favorable for hydration.

Length of curing period. The length of time that concrete should be protected against loss of moisture is depedent upon the type of cement, mix proportions, required strength, size and shape of the concrete mass, weather, and future exposure conditions. The period may vary from a few days to a month or longer. The influence of curing on the strength of concrete is shown in figure 5–20.

Curing Methods

Concrete can be kept moist, and in some cases at a favorable temperature by a number of curing methods. These may be classified into two categories: methods that supply additional moisture, and methods that prevent loss of moisture. Table 5–4 lists several effective methods of curing concrete together with comparable advantages and disadvantages.

Methods that Supply Additional Moisture. These methods add moisture to the surface of the concrete during the early hardening period. Such methods include sprinkling and wet coverings. Some cooling, through evaporation, is provided which is beneficial in hot weather.

Sprinkling. Continuous sprinkling with water is an excellent method of curing. If sprinkling is done at intervals, the concrete must not be allowed to dry out between applications. The ex-

Table 5-4. Curing Methods

Method	Advantage	Disadvantage
Sprinkling with water or covering with wet burlap.	Excellent results if constantly kept wet.	Likelihood of drying between sprinklings. Difficult on vertical walls.
Straw	Insulator in winter	Can dry out, blow away, or burn.
Moist earth	Cheap, but messy	Stains concrete. Can dry out. Removal problem.
Ponding on flat surfaces	Excellent results, maintains uniform temperature.	Requires considerable labor, undesirable in freezing weather.
Curing compounds	Easy to apply. Inexpensive	Sprayer needed. Inadequate coverage allows drying out. Film can be broken or tracked off before curing is completed. Unless pigmented, can allow concrete to get too hot.
Waterproof paper	Excellent protection, prevents drying	Heavy cost can be excessive. Must be kept in rolls, storage and handling problem.
Plastic film	Absolutely watertight, excellent protection. Light and easy to handle.	Should be pigmented for heat protection. Requires reasonable care and tears must be patched. Must be weighed down to prevent blowing away.

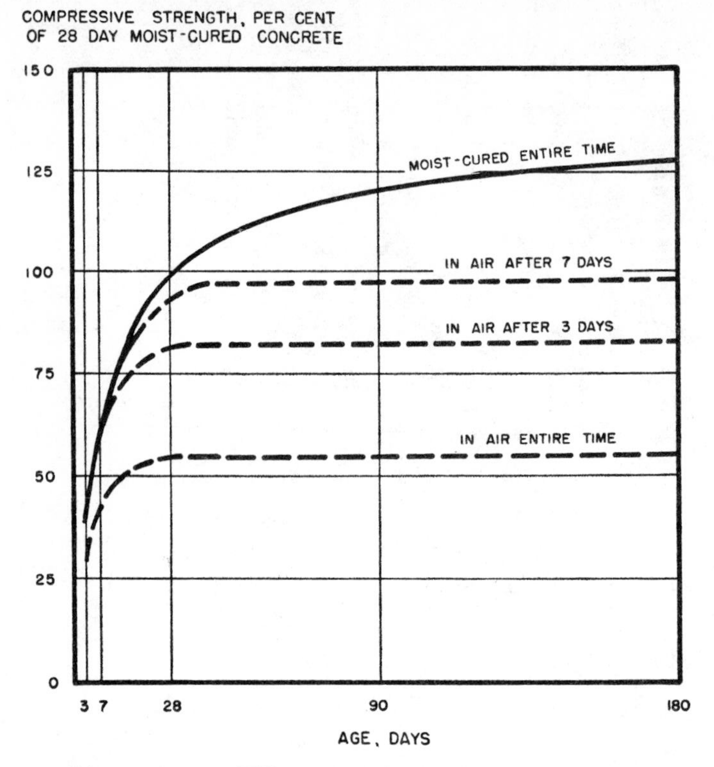

Figure 5–20. Effect of moist curing on concrete compressive strength.

pense involved and volume of water required may be disadvantageous.

Wet coverings. Wet coverings such as burlap, cotton mats, and other moisture retaining fabrics are used extensively for curing. Straw and moist earth may also be used. These coverings should be placed as soon as the concrete has hardened sufficiently to prevent surface damage. Wet coverings should remain in place and kept moist during the entire curing period. For most structural use, the curing period for cast in place concrete is usually 3 days to 2 weeks, depending upon such conditions as temperature, cement type, mix proportions, etc. More extended curing periods are desirable for bridge decks and other slabs exposed to weather and chemical attack.

Methods that Prevent Loss of Moisture. These methods prevent loss of moisture by sealing the surface. This may be done by means of waterproof paper, plastic film, liquid-membrane-forming compounds, and forms left in place.

Waterproof paper. Waterproof paper is an efficient means of curing horizontal surfaces and structural concrete of relatively simple shapes. The paper should be large enough to cover the width and edges of the slab. Adjacent sheets should be lapped 12 inches or more and the edges should be weighted down, to form a continuous cover with completely closed joints. The surface should be wet with a fine spray of water before covering. The coverings should remain in place during the entire curing period.

Plastic film. Certain plastic sheet materials are used in curing concrete. They are lightweight, effective moisture barriers and are easily applied to either simple or complex shapes. In some cases, thin plastic sheets may discolor hardened concrete, especially if the surface has been steel-troweled to a hard finish. Coverage, overlap, weighting down of edges, and surface wetting requirements are similar to those for waterproof paper.

Curing compounds. Curing compounds retard or prevent evaporation of moisture from the concrete. They are suitable not only for curing fresh concrete, but also for further curing of concrete after removal of forms or after initial moist curing. They are applied by spray equipment. Curing compound can be satisfactorily applied with hand operated pressure sprayers on odd widths or shapes of slabs and on concrete surfaces exposed by the removal of forms. Concrete surfaces subjected to heavy rain within 3 hours after curing compound is applied should be resprayed. Brushes may be used to apply the compound on formed surfaces, but should not be used on unformed concrete due to the danger of marring the concrete, opening the surface to excessive penetration of the compound, and breaking the continuity of the film. Curing compounds permit curing to continue for long periods while the concrete is in use. Curing compounds may prevent bond between hardened and fresh concrete, consequently they should not be used if bond is necessary.

(4) *Forms left in place.* Forms can provide adequate protection against loss of moisture if the top exposed concrete surfaces are kept wet. Wood forms left in place should be kept moist by sprinkling, especially during hot, dry weather.

EFFECTS OF TEMPERATURE

Hot Weather Concreting

Problems. Concreting in hot weather poses some special problems, among which are reduction in strength and cracking of flat surfaces due to rapid drying. Concrete may stiffen before it can be consolidated because of rapid setting of the cement and excessive absorption and evaporation of mixing water. This leads to difficulty in the finishing of flat surfaces. During hot weather, precautions should be taken to limit concrete temperature to less than 90°. Limitations are imposed on the maximum temperature of concrete and on the placing of concrete during hot weather because the quality and durability suffer when concrete is mixed, placed, and cured at high temperatures. Difficulty can be experienced even with concrete temperatures of less than 90° F. The combination of hot, dry weather and high winds is most severe, especially when placing large exposed slabs.

Effects of High Concrete Temperatures.

Mixing water requirements. High temperatures accelerate the hardening of concrete and more mixing water is generally required for the same consistency. Figure 5–21 shows the increase in mixing water required to maintain the same slump as temperature increases; however, increasing the water content of concrete without increasing the cement content results in a higher water-cement ratio, thereby adversely affecting the strength and other properties of hardened concrete.

Compressive strength of concrete. Figure 5–22 shows the effect of high concrete temperatures on compressive strength. These tests, using identical concretes of the same water-cement ratio, show that while higher concrete temperatures increase early strength, at later ages the re-

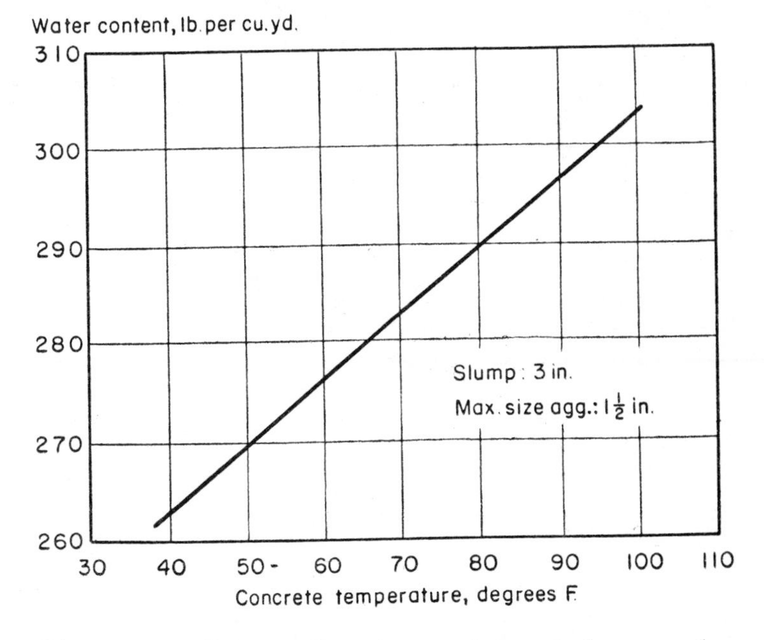

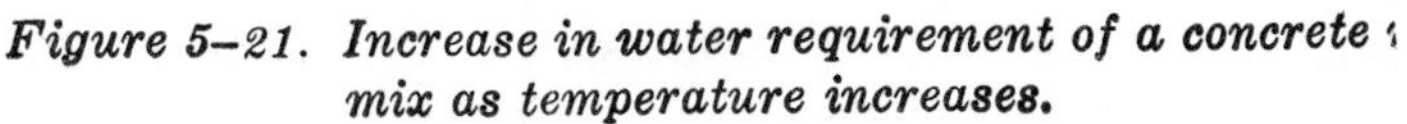

Figure 5–21. Increase in water requirement of a concrete mix as temperature increases.

verse is true. If the water content had been increased to maintain the same slump (without changing the cement content), the reduction in strength would have been even greater than shown in figure 5–22.

Cracking. In hot weather the tendency for cracks to form is increased both before and after hardening. Rapid evaporation of water from hot concrete may cause plastic shrinkage cracks before the surface has hardened. Cracks may also develop in the hardened concrete because of increased shrinkage due to higher water requirement and because of the greater range between

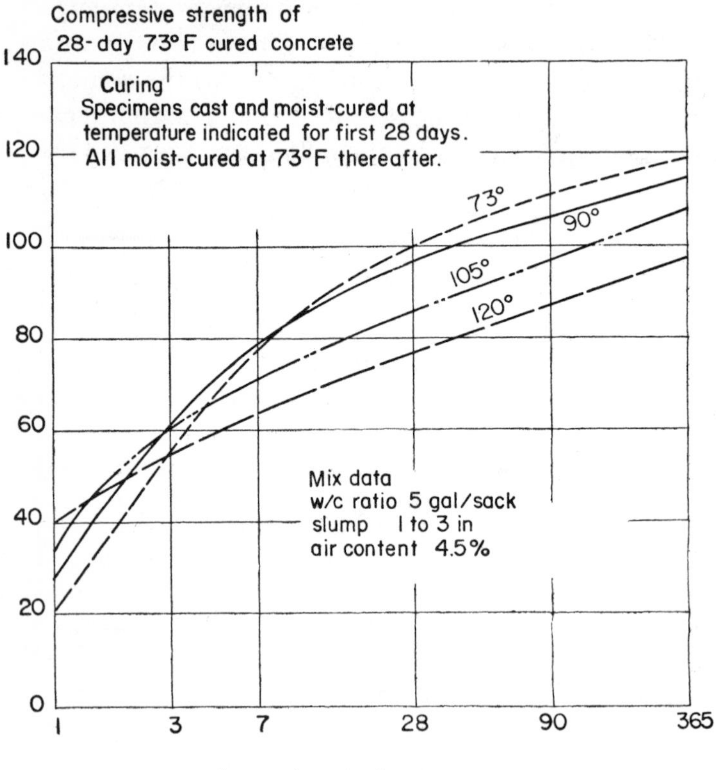

Figure 5–22. Effect of high temperature on concrete compressive strength at various ages.

the high temperature at the time of hardening and the low temperature to which the concrete will later drop.

Cooling Concrete Materials. The most practical method of maintaining low concrete temperatures is to control the temperature of the concrete materials. One or more of the ingredients may be cooled before mixing. In hot weather the aggregates and water should be kept as cool as practicable. Mixing water is the easiest to cool and is the most effective, pound for pound, for lowering the temperature of concrete. However, since aggregates represent 60 to 80 percent of the total weight of concrete, the concrete temperature is primarily dependent on the aggregate temperature. Figure 5–23 shows the effect of the temperature of mixing water and aggregate on the temperature of fresh concrete. The temperature of the fresh concrete can be lowered by several means:

(1) Using cold mixing water. Slush ice can be used in extreme cases to cool the water.

(2) Cooling coarse aggregate by sprinkling; avoid excessive use of water.

(3) Insulating mixer drums or cooling them with sprays or wet burlap coverings.

(4) Insulating water-supply lines and tanks, or painting them white.

(5) Shading materials and facilities not otherwise protected from the heat.

(6) Working only at night.

(7) Avoiding the use of hot cement.

(8) Sprinkling forms, reinforcing steel, and subgrade with cool water just before placing concrete.

Additional Precautions. High temperatures increase the rate of concrete hardening and

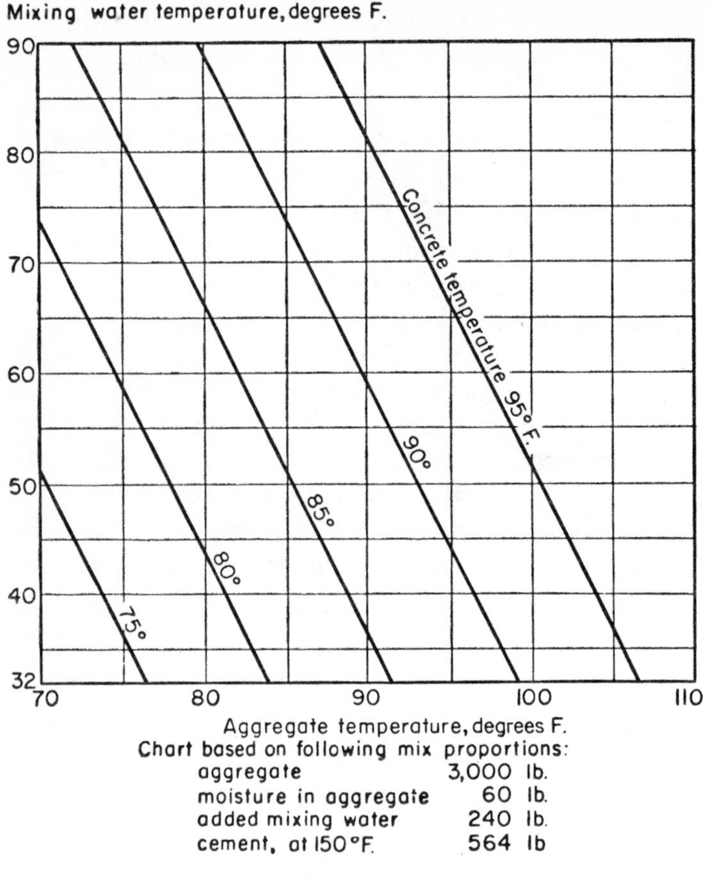

Figure 5–23. Temperature of fresh concrete as affected by temperature of materials.

shorten the length of time within which the concrete can be handled and finished. Since setting is accelerated, transporting and placing should be done as quickly as practicable. Extra care must be taken in placing techniques to avoid cold joints. Curing is difficult in hot weather since the water evaporates rapidly from the concrete. Proper curing is especially important in hot weather because of the greater danger of crazing and cracking. Forms are not satisfactory substitutes for curing in hot weather. They should be loosened as soon as it can be done without damage to the concrete. Water should be applied and allowed to cover the concrete. The efficiency of curing compounds is reduced in hot weather. Frequent sprinkling and the use of wet burlap and other means of retaining the moisture for longer periods are necessary.

Cold Weather Concreting

Considerations. The placing of concrete does not have to be suspended during winter months if necessary precautions are taken to protect the fresh concrete from freezing temperatures until necessary protection can be provided. For successful winter work, adequate protection must be provided when temperatures of 40° F. or lower occur during placing and during the early curing period. Prior planning should include provisions for heating concrete materials and maintaining favorable temperatures after the concrete is placed. To prevent freezing, the temperature of the concrete should not be less than that shown in line 4, table 5–5, at the time of placing. Thermal protection may be required to assure that subsequent concrete temperatures do not fall below the minimum shown in line 5, table 5–5, for the periods shown in table 5–6, to ensure durability or to develop

*Table 5–5. Recommended Concrete Temperatures for Cold Weather Construction**

CONDITION/CURING	TEMPERATURE, °F		
	THIN SECTIONS	MOD. SECTIONS	MASS SECTIONS
AS-MIXED FRESH CONCRETE:			
1. Above 30°F	60	55	50
2. 0 – 30°F	65	60	55
3. Below 30°F	70	65	60
AS-PLACED FRESH CONCRETE:			
4. Low ambient	55	50	45
5. Max. gradual drop in first 24 hr after end of protection.	50	40	30

strength. The temperature of fresh concrete as mixed should not be less than shown in lines 1, 2, and 3 of table 5–5. Note that lower concrete temperatures are recommended for heavy mass sections than for thinner sections because less heat is dissipated during the hydration period. More heat is lost during transporting and placing; therefore, the fresh concrete temperatures are higher for colder weather. Temperatures for concrete over 70° F. are seldom necessary, since they do not furnish proportionately longer protection from freezing because the loss of heat is greater. High concrete temperatures also require more mixing water for the same slump; this contributes to cracking due to shrinkage.

Effect of Low Concrete Temperatures. Temperature affects the rate at which hydration of cement occurs—low temperatures retard concrete hardening and strength gain. This fact is illustrated in figure 5–24 for concrete mixed, placed and cured at tempeatures between 40° and 73° F. It can be seen from the above figure that at temperatures below 73° F., the strength of concrete is lower during the first 28 days and then is higher than that of concrete cured at 73° F. Therefore, concrete placed at temperatures below 73° F. must be cured longer. It should be remembered that

Table 5–6. Recommended Duration of Protection for Concret Placed in Cold Weather (Air-entrained Concrete)*

Degree of exposure to freeze-thaw	Normal concrete**	High-early-strength concrete†
No exposure	2 days	1 day
Any exposure	3 days	2 days

*Protection for durability at temperature indicated in line 4, Table 5–5. Adapted from Recommended Practice for Cold Weather Concreting (ACI 306–66).

**Made with Type I, II, or Normal cement.

†Made with Type III or High-Early-Strength cement, or an accelerator, or an extra 100 lb. of cement.

strength gain practically stops when moisture required for hydration is no longer available. The early strengths that may be achieved through use of Type III or High-Early Strength cement are higher than those achieved with Type I cement, as illustrated by figure 5–25.

Cold Weather Techniques.

Heating concrete materials. Thawing frozen aggregates makes proper batching easier. Frozen aggregates must be thawed to avoid pockets of aggregate in the concrete after placing. Excessively high water content must be avoided if thawing takes place in the mixer. It is seldom necessary to heat aggregates in temperatures above

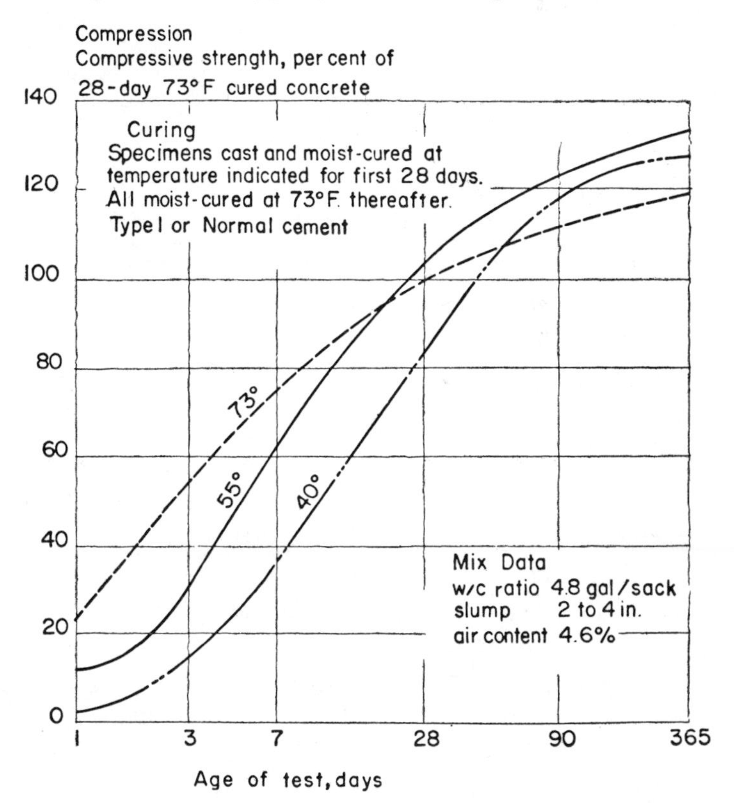

Figure 5–24. Effect of low temperatures on concrete compressive strength at various ages.

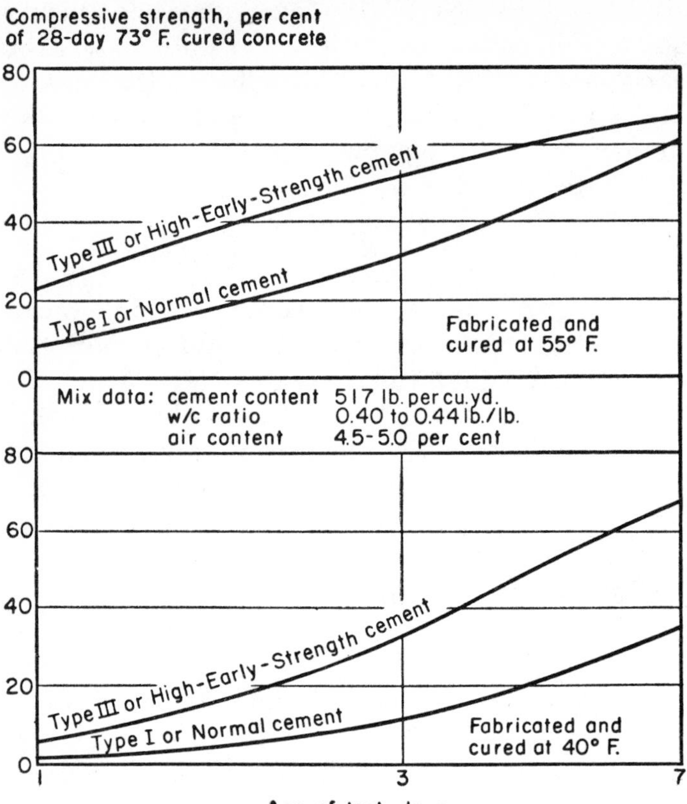

Figure 5–25. Early compressive strength relationships involving portland cement types and low curing temperatures.

freezing. At temperatures below freezing, concrete of the required temperatures may be produced by heating the fine aggregate only.

Heating aggregates. Several methods are used to heat aggregates. On small jobs aggregates may be heated by piling them over metal pipes in which fires are built. To obtain recommended concrete temperatures, the average temperature of the aggregates should not exceed 150° F. Aggregates are frequently stockpiled over pipes through which steam is circulated. The stockpiles should be covered with tarpaulines to retain and distribute the heat. Live steam may be injected directly into the pile of aggregate to heat it but the problem of variable moisture content can result in erratic control of mixing water.

Heating water. Mixing water is the easiest ingredient of concrete to heat. It can store five times as much heat as solid materials of the same weight, although the weight of aggregates and cement is much greater than the weight of water. The heat stored in water may be used to heat concrete materials. When either aggregates or water are heated above 100° F., they should be combined in the mixer before the cement is added. Figure 5–26 shows the effect of the temperature of materials on the temperature of fresh concrete. Figure 5–26 is reasonably accurate for most ordinary concrete mixtures. Water should not be hotter than 180° F. shown on the chart. The chart further indicates that in some cases both aggregates and water must be heated. For example, if the weighted average temperature of aggregates is below 36° F. and the desired concrete temperature is 70° F., then the aggregates would have to be heated in order to limit the water temperature to 180° F.

Use of high-early-strength concrete. High strength at an early age is frequently desired during winter construction to reduce the length of time protection is required. The value of high-early-strength concrete during cold weather is often realized through early re-use of forms and removal of shores, savings in the cost of additional heating and protection, earlier finishing of flatwork, and earlier use of the structure.

Use of accelerators. So called antifreeze compounds or other materials should not be used to lower the freezing points of concrete. Strength of concrete and other properties are seriously affected by the large quantities of an accelerator required to lower the freezing point appreciably. Small amounts of additional cement or of accelerators such as calcium chloride to accelerate hardening of concrete in cold weather may be beneficial,

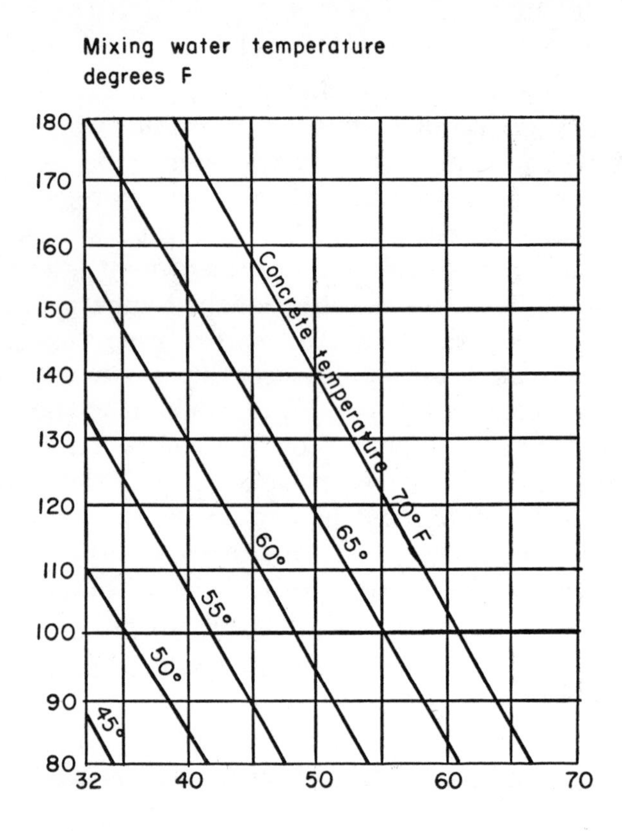

Chart based on following mix proportions

aggregate	3,000 lb
moisture in aggregate	60 lb
added mixing water	240 lb
cement	564 lb

Figure 5–26. Temperature of mixing water needed to produce heated concrete of required temperature.

as long as no more than 2 percent of calcium chloride by weight of cement is used. Precautions are necessary when using accelerators containing chlorides where there is an in-service potential for corrosion, as, for example, for prestressed concrete, or where aluminum inserts are contemplated. When sulfate-resisting concrete is required, use an extra sack of cement per cubic yard rather than calcium chloride. Accelerators should-not be used as a substitute for proper curing and frost protection.

Preparation before placing. Concrete should never be placed on a frozen subgrade. Severe cracks usually occur due to settlement when the subgrade thaws. If the subgrade is frozen for only a few inches deep, the surface may be thawed by burning straw, by steaming, or if the grade permits, by spreading a layer of hot sand or other granular material. The ground must be thawed enough to assure that it will not freeze during the curing period.

Curing. Concrete in forms or covered with insulation seldom loses enough moisture at 40° to 55°F. to impair curing. Some moisture must be provided for concrete curing during winter to offset the drying tendency when heated enclosures are used. Concrete should be kept at a favorable temperature until it is strong enough to withstand low temperatures and anticipated service loads. Forms serve to distribute the heat more evenly and help prevent drying and overheating. They should be left in place as long as practicable. Concrete that is allowed to freeze soon after placing is permanently damaged. If the concrete has been frozen once at an early age, it may be restored to nearly normal under favorable curing conditions although it will not weather as well nor be as watertight as concrete that is not frozen. Air-entrained concrete is less susceptible to damage from freezing than concrete without entrained air. Cold weather concreting is discussed in detail in TM 5–349. Several methods to maintain proper temperatures during curing are described below.

Live steam. Live steam exhausted into an enclosure is an excellent practical method of curing in extremely cold weather because moisture from the steam offsets the rapid drying that occurs when very cold air is heated. A curing compound may be used after the protection is removed and the air temperature is above freezing.

Insulation blanket or bat insulation. The manufacturers of these materials can usually provide information on the amount of insulation necessary for protection at various temperatures. The corners and edges of concrete are most vulnerable to freezing and should be checked to determine the effectiveness of the protective covering.

(3) *Heated enclosures.* Wood, canvas, building board, plastic sheets, or other materials are used to enclose and protect concrete from below freezing temperatures. Wood framework covered with tarpaulins or plastic sheets is also used. The enclosures should be sturdy and reasonably airtight. Free circulation of warm air should be provided for. Control of the temperature within the enclosure is easiest with live steam. Carbon-dioxide-producing heaters (salamanders and other fuel-burning heaters) should not be used during concrete placing and for 24 to 36 hours after placement unless they are properly vented. Temperatures differences should be minimized. Adequate minimum temperatures should be provided for the entire curing period.

FORM REMOVAL

It is generally advantageous to leave forms in place throughout the required curing period. However, it may be necessary to strip forms as early as possible to permit their immediate reuse. Also certain finishing operations, such as rubbing, may require early removal of forms. In any case, forms must not be removed before the concrete is strong enough to carry its own weight and any other loads that may be placed on it during construction. The forms for columns, footings, and sides of beams and walls can usually be removed before the forms for floors and beam bottoms. For most conditions, it is better to rely on the strength of the concrete as determined by test rather than to select arbitrarily the age at which forms may be removed. A minimum compressive strength of 500 psi should be attained before concrete is exposed to freezing. The age-strength relationship should be determined from tests on representative samples. Under average conditions (for example, air-entrained concrete made with a water-cement ratio of 6 gallons per sack) the times required to attain certain strengths are shown in table 5–7. It should be remembered that strengths are affected by materials used, temperatures, and other condi-

*Table 5–7. Age-Strength Relationship**

Strength, psi	Age	
	Type I or normal	Type III or high-early-strength
500	24 hours	12 hours
750	1½ days	18 hours
1,500	3½ days	1½ days
2,000	5½ days	2½ days

*Air-entrained concrete with water-cement ratio of 6 gallons per sack.

tions. The time required before form removal, therefore, will vary from job to job.

Form Removal Procedures

Forms should be designed and constructed with some thought as to their removal with a minimum of danger to the concrete. The forms must be stripped carefully to avoid damage to the surface of the concrete. When it is necessary to wedge against the concrete, only wood wedges should be used rather than a pinchbar or other metal tool. The forms should not be jerked off after wedging has been started at one end; this is almost certain to break the edges of the concrete. Forms that are to be reused should be cleaned and oiled immediately after their removal. Nails should be withdrawn as the forms are stripped from the concrete.

PATCHING

New Concrete

Inspection. Concrete should be inspected for surface defects when the forms are removed. These defects may be rock pockets, inferior quality, ridges at form joints, bulges, bolt holes, and form-stripping damage. Repairs are costly and interfere with the use of the structure. However, experience has demonstrated that no step in the procedure can be omitted or carelessly performed without harming the service ability of the repair work. If not properly performed, the repair will later become loose, will crack at the edges, and will not be water-tight.

Timely repair. On new work the repairs which will develop the best bond and thus have the best chance of being as durable and permanent as the original work are those made immediately after early stripping of the forms, while the concrete is quite green. For this reason, repairs should be performed within 24 hours after the forms have been removed.

Removal of ridges and bulges. If ridges and bulges are objectionable, they may be removed by careful chipping followed by rubbing with a grinding stone.

Patching Concrete.

Defective areas. Defective areas such as rock pockets or honeycomb must be chipped out to solid concrete, the edges cut as straight as possible at right angles to the surface or slightly undercut to provide a key at the edge of the patch. The surface of all holes that are to be patched should be kept moist for several hours before applying the mortar. The mortar should be allowed to set as long as possible before being used, to reduce the

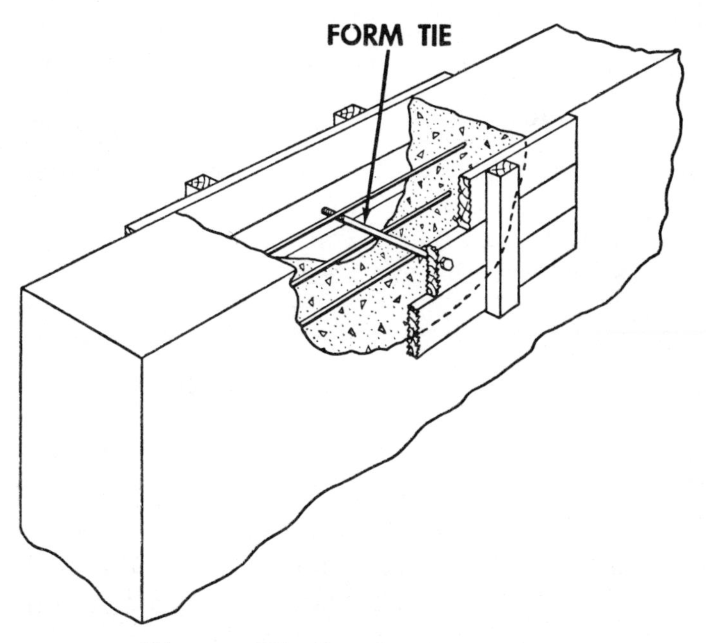

Figure 5–27. Repair of new concrete.

amount of shrinkage and make a better patch. If a shallow layer of mortar is placed on top of the honeycombed concrete, moisture will form in the voids and subsequent weathering will cause the mortar to spall off. Shallow patches may be filled with mortar placed in layers not more than ½ inch thick. Each layer should be scratched rough to improve the bond with the succeeding layer, and the last layer smoothed to match the adjacent surface. Where absorptive form lining has been used, the patch can be made to match the rest of the surface by pressing a piece of the form lining against the fresh patch.

Large patches. Large or deep patches may be filled with concrete held in place by forms. These patches should be reinforced and doweled to the hardened concrete (fig. 5–27). Patches usually appear darker than the surrounding concrete. Some white cement should be used in the mortar or concrete used for patching if appearance is important. A trial mix should be tried to determine the best proportion of white and gray cements to use. Before mortar or concrete is placed in patches, the surrounding concrete should be kept wet for several hours. A grout of cement and water mixed to a creamy consistency should then be brushed into the surfaces to which the new material is to be bonded. Curing should be started as soon as possible to avoid early drying. Damp burlap, tarpaulins and membrane curing compounds are useful for this purpose.

Filling bolt holes. Bolt holes should be filled with mortar carefully packed into place in small amounts. The mortar should be mixed as dry as possible, with just enough water so that it will be tightly compacted when forced into place. Tie rod holes extending through the concrete can be

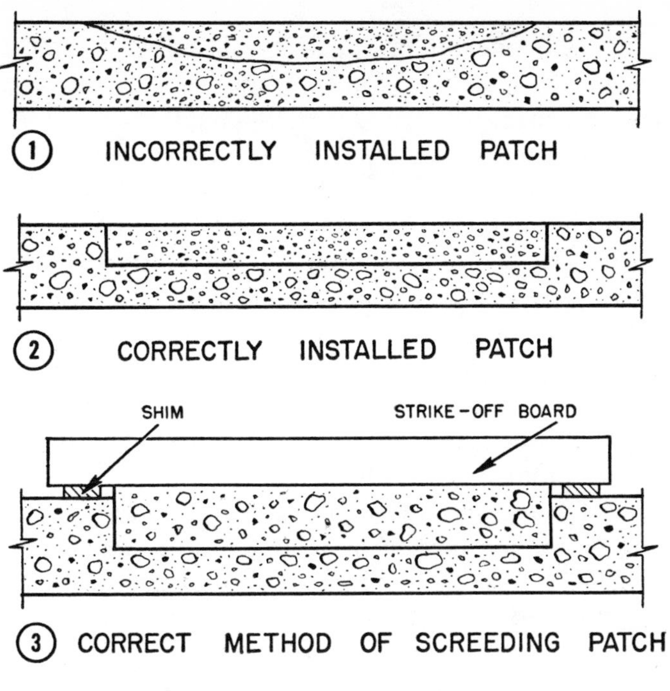

Figure 5–28. Patching concrete.

filled with mortar with a pressure gun similar to an automatic grease gun.

Flat surfaces. Feathered edges around a patch (1 fig. 5–28) will break down. The chipped area should be at least 1 inch deep with the edges at right angles to the surface (2 fig. 5–30). The correct method of screeding a patch is shown in 3 figure 5–28. The new concrete should project slightly beyond the surface of the old concrete. It should be allowed to stiffen and then troweled and finished to match the adjoining surfaces.

Old Concrete

Inspection. Before repairing old concrete, the amount of material to be removed must first be determined. A thorough inspection of the imperfection should be made before repairs are started. All concrete of questionable quality should be removed. However, in many cases, if all of the weakened material were removed, nothing would be left. In the event that all the weakened material cannot be removed and the old concrete will be completely encased in new concrete, only the loose material needs to be removed. Where old and new concrete form a junction at a surface exposed to weathering or chemical attack, the old concrete must be perfectly sound. It is far better to remove too much old concrete than too little.

Preparation. After initially removing the weakened material and loose particles, the surface to be repaired should be thoroughly cleaned with air or water, or both. The area around the repair should be kept continuously wet for several hours, preferably overnight. This wetting is especially important in the repair of old concrete. Without wetting a good bond cannot be achieved.

Patching Concrete. Where small areas of patching are involved, rectangular patches should be used. The upper 1 to 2 inches of the edge of the old concrete should be trimmed to a vertical face to eliminate the possibility of thin edges in the patch or in the old concrete. The depth of the repair is dependent upon many conditions. For repairing large structures such as walls, piers, curbs, and slabs, the depth of repair should be at least 6 inches where possible. If reinforcement bars are in the old concrete, there should be a clearance of at least an inch around each exposed bar. After the wetting period the new concrete should be placed in layers and each layer thoroughly tamped. The concrete should be a low-slump mixture which has been allowed to stand for a while in order to reduce shrinkage in the hole. In the repair of old concrete, it may be necessary to use forms to hold the new concrete in place. The design and construction of these forms often calls for a high degree of ingenuity. Well designed and properly constructed forms are important steps in the procedure for repairing concrete. Deep patches should be reinforced and to the hardened concrete. Following patching, good curing is essential. Curing should be started as soon as possible to avoid early drying.

6

Reinforced Concrete Construction

DEVELOPMENT

Reinforced Concrete. Concrete is strong in compression, but relatively weak in tension. The reverse is true for slender steel bars and when the two materials are used together one makes up for the deficiency of the other. When steel is embedded in concrete in a manner which assists it in carrying imposed loads, the combination is known as reinforced concrete. Beam strength can be increased significantly by the use of steel in the tension side (fig. 6–2).

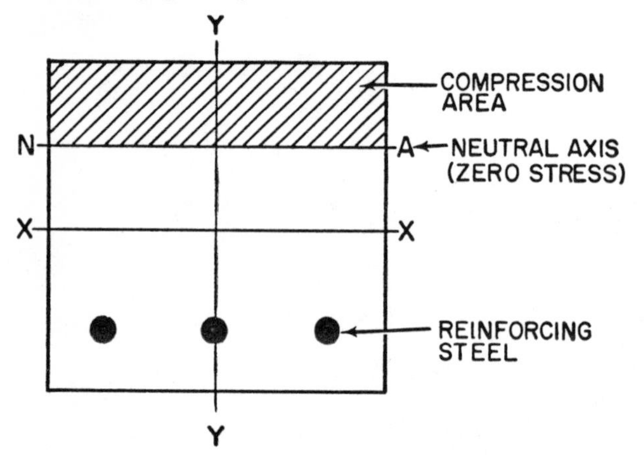

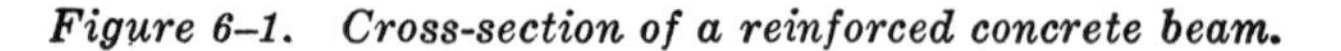
Figure 6–1. Cross-section of a reinforced concrete beam.

Tensile Strength. Tensile strength is such a small percentage of the compressive strength that it is ignored in reinforced concrete beam calculations. Instead, tensile resistance is provided by longitudinal steel bars well embedded in the tension side.

Shear Strength. The shear strength of concrete is about one-third the unit compressive strength, and tensile strength is less than one-half the shear strength. A concrete slab subjected to a downward concentrated load fails due to the diagonal tension. Beams are prevented from failing in diagonal tension by providing web reinforcement.

Bond Strength. Bond strength is the resistance developed by concrete to the pulling out of a steel bar embedded therein. The theory of reinforced concrete beam design is based on the assumption that a bond exists between the steel and concrete which prevents relative movement between them as the load is applied. The amount of bond strength that can be developed depends largely upon the area of contact between the two materials. Due to their superior bond value, bars manufactured with a very rough outside surface, called deformed bars (fig. 6–2), have replaced plain bars.

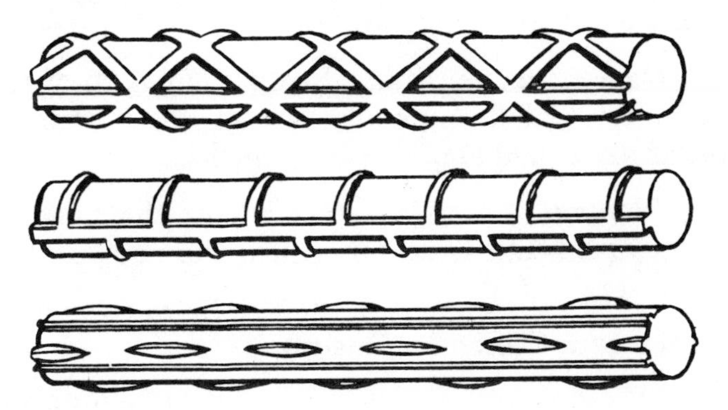

Figure 6–2. Steel reinforcing bars.

Bending Strength. When a beam is subjected to a bending moment it deflects because the parts that are in compression shorten and those that are in tension become longer. The weak portions of the beam, as shown by the short irregular lines in figure 6–3 where tension exists must be reinforced with steel. This figure is not intended to show that beams always crack excessively, but to show the condition that beams may reach if they are loaded sufficiently. Concrete in the areas subjected to compression is usually effective by itself.

Creep. The tendency for loaded concrete to deform after a lapse of time is known as creep or plastic flow. Concrete tends to exhibit this continuing deformation over the whole stress range. It takes place rapidly at first, then much more slowly, becoming small or negligible after a year or two. Some authorities believe that the initial strains of well-designed reinforced concrete structures are removed during the first few service loadings and after that they perform elastically as long as they are not overloaded. Due to creep, deflection of concrete cannot be predicted by the common deflection formulas with any satisfactory degree of accuracy. However, failures are not usually traceable to creep because this phenomenon usually ceases in well proportioned structures before excessive deflections occur.

Homogeneous Beams. Beams composed of the same material throughout, such as steel or timber beams are called homogeneous beams.

Neutral Axis. The axis where the bending stress in a beam is zero is called the neutral axis (fig. 6–1).

Reinforced Concrete Design

Specifications. The solutions to common problems of reinforced concrete design are influenced

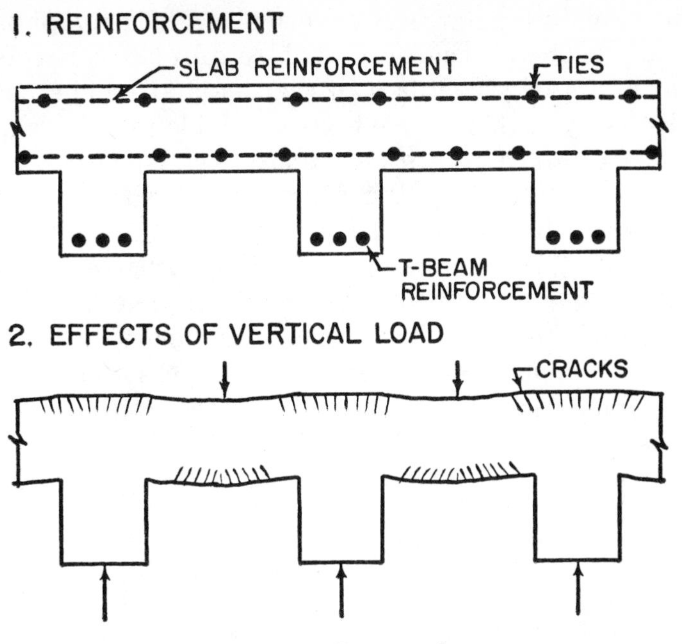

Figure 6–3. Concrete beams.

by the practical experience of many structural engineers and by the results of exhaustive tests and investigations conducted at universities and elsewhere. The results of these experiences, tests, and investigations have been reflected in rules and methods that are published as the "Report of the Joint Committee on Standard Specifications for Concrete and Reinforced Concrete" and other references. In most practical designs engineers will make reference to standard specifications.

Design. The term "design of a beam" for instance denotes determination of the size and the materials required to contract a beam that can safely support specified loads under certain definite conditions of span, stress, and the like. Economy and efficiency in the use of materials, strength, spacing, and arrangement of reinforcing steel are factors which enter into the design. The design of a reinforced concrete structure requires sound engineering judgment and experience. No attempt is made to teach the design of reinforced concrete members or structures in this manual in view of the many authoritative texts available in this field. The design of reinforced concrete consists principally in predicting the position and direction of potential tension cracks in concrete, and in forestalling the cracking by locating sufficient steel across them. From the structural engineer's point of view, there are three types of members, namely, tension members, compression members and bending members, called beams. Beams require the most study because the bending stress varies over the cross section, instead of being uniformly distributed.

Workmanship. It is emphasized that the best of designs can be ruined if the intent of the plans is not carried out faithfully and intelligently in the field. Proper reinforced concrete construction depends on men who understand the action of structures and who appreciate the characteristics and limitations of the material.

STRUCTURAL MEMBERS

Reinforcing

Types of Structural Members. A reinforced concrete structure is made up of many types of reinforced structural members including columns, beams, girders, walls, footings, slabs, etc. Analysis indicates that the different members interact to a considerable degree in view of the fact that a reinforced concrete structure is monolithic.

Beam Reinforcement. Four common types of beam reinforcing steel are shown in figure 6–4. Both straight and bent-up principal reinforcing bars are depended on to resist the bending tension in the bottom over the central portion of the span. Fewer bars are necessary on the bottom near the ends of the span where the bending moment is small. For this reason, some bars may be bent as shown in figure 6–4 so the inclined portion can be used to resist diagonal tension. The reinforcing bars of continuous beams are continued across the supports to resist tension in the top in that area.

When there are not enough bent bars available to resist all the diagonal tension, additional U-shaped bars, called stirrups, are usually necessary. Due to the tensile stress on the stirrups, they must pass under the bottom steel and perpendicular to it to prevent lateral slippage. Welded stirrups serve the same purpose as U-shaped bars and may be placed at any desired angle decided on by the engineer.

Horizontal reinforcing steel is usually supported on devices called bolsters or chairs (fig. 6–5) that hold the bars in place during construction. Stress-carrying reinforcing bars must be placed in accordance with American Concrete Institute (ACI) Code 318–63, section 808, Building Code Requirements for Reinforced Concrete.

Column Reinforcement. A column is a slender, vertical member which carries a superimposed load. Concrete columns must always be reinforced with steel, unless the height is less than

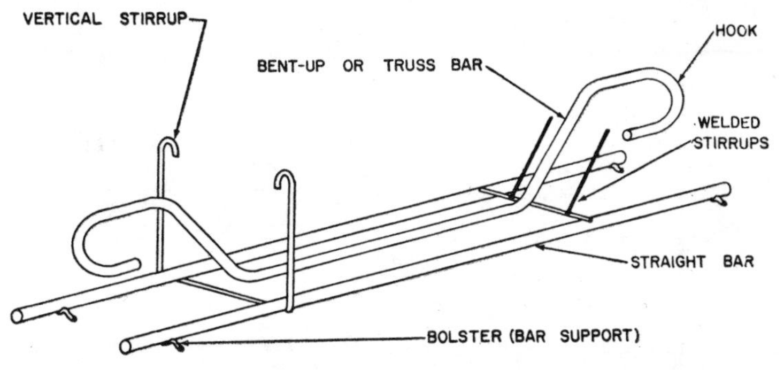

Figure 6–4. Typical shapes of reinforcing steel.

three times the least lateral dimension in which case the member is called a pier or pedestal. Allowable loads and minimum column dimensions are governed by ACI Code 318–63. Most concrete columns are subjected to bending. Figure 6–6 shows two types of column reinforcement. Vertical reinforcement is the principal reinforcement. Lateral reinforcement surrounds the column horizontally and consists of individual ties ((a), fig. 6–6) or a continuous spiral ((b), fig. 6–6).

Tied Columns. A loaded concrete column shortens vertically and expands laterally ((a), fig. 6–6). Lateral reinforcement in the form of lateral ties is used to restrain the expansion. The principal value of lateral reinforcement is to provide intermediate lateral support for the vertical, or longitudinal reinforcement. Columns reinforced in this manner are called tied columns.

Spiral Columns. A spiral column is identified by a continuous spiral winding ((b), fig. 6–6) which encircles the core and longitudinal steel. A spiral column is generally considered to be more substantial than a tied column due to the continu-

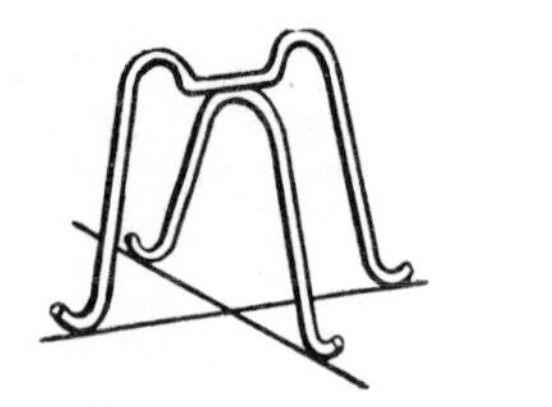

HIGH CHAIR—HC

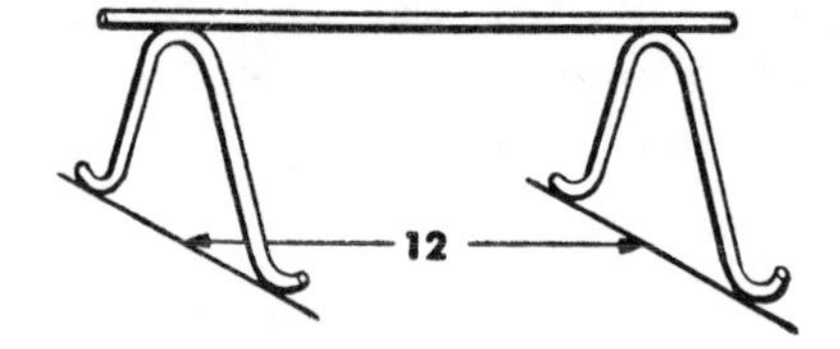

CONTINUOUS HIGH CHAIR—CHC

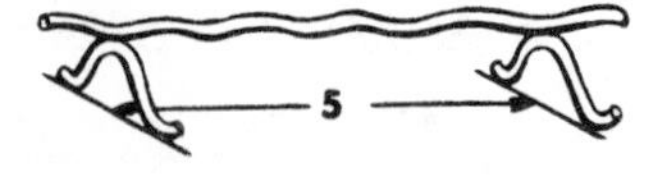

SLAB BOLSTER—SB

BEAM BOLSTER—BB

Figure 6–5. Supports for reinforcing steel.

ity of the spiral reinforcement, as opposed to the many imperfect anchorages at the ends of the individual lateral ties in a tied column. The pitch of the spiral reinforcement can be reduced to provide effective lateral support. The pitch of the spiral and tie size and number of bars are specified by the engineer.

Composite and Combination Columns. A structural steel or cast iron column thoroughly encased in concrete, reinforced with both longitudinal and spiral reinforcement is called a composite column. The cross-sectional area of the metal core of a composite column cannot exceed 20 percent of the gross area of the column. A structural steel column encased in concrete at least 2½ inches thick over all the metal and reinforced with welded wire fabric is called a combination column. Composite and combination columns are often used in construction of large buildings.

Vertical Reinforcement. The vertical reinforcement in a column helps to carry the direct axial load as the column shortens under load. Vertical bars are located around the periphery of a column for effective resistance to possible bending. Each vertical reinforcing bar tends to buckle outward in the direction of least opposition. For this reason every vertical reinforcing bar should be held securely, at close vertical intervals against outward lateral movement. For example in the 8-bar group shown in (b), figure 6–6, a second system of ties, T_2 is necessary to confine the four intermediate vertical reinforcing bars. If these ties are omitted, the 8-bar group tends to come into a slightly circular configuration under load. The resultant bulging leads to destructive cracking of the concrete shell and failure of the column. A round column has obvious advantages in this respect.

Shrinkage and Temperature Reinforcement

Slabs and walls must not only be reinforced by the principal reinforcement against the applied loads to which they are subjected. They must also be reinforced in the lateral direction to resist the effects of shrinkage and temperature change. Concrete shrinks as hydration proceeds. A small percentage of steel must be used to resist this force. Similarly, concrete must be allowed to contract

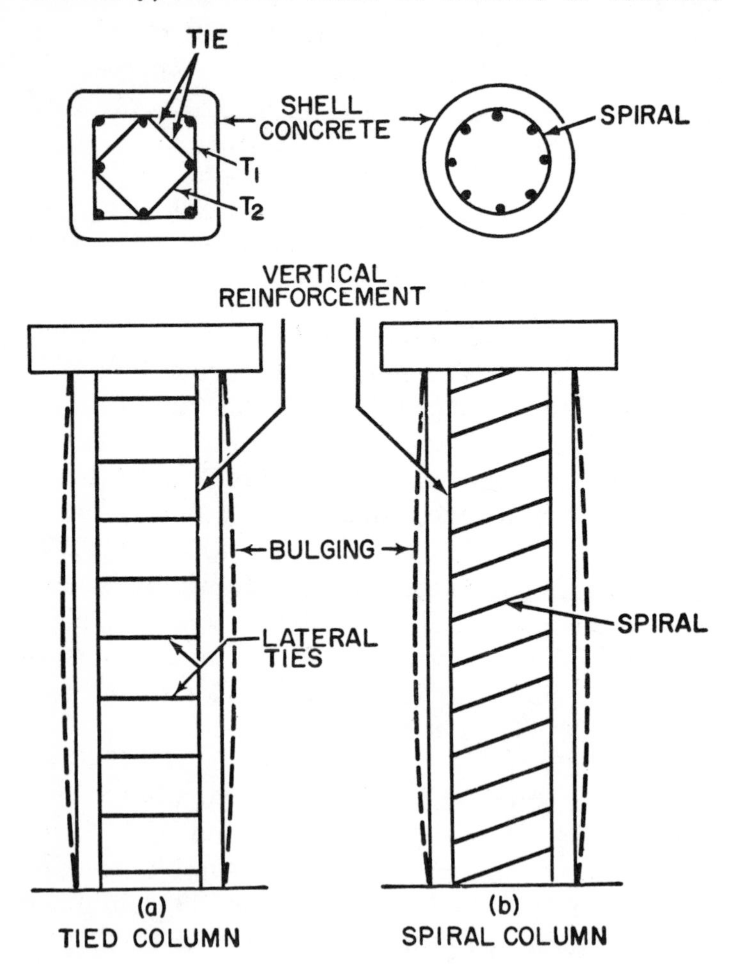

Figure 6–6. Reinforced concrete columns.

with the lowering of temperature. A fall in temperature of about 83 degrees causes as much movement as drying shrinkage. Depending on the extent adjacent construction interferes, this movement tends to cause the concrete to become stressed in tension. The amount of shrinkage and temperature reinforcement usually provided for is approximately one-fifth of 1 percent of the area of the cross section of concrete, as required by the specifications.

REINFORCING STEEL

Grades. Concrete reinforcing steel is available as bars in 11 bar sizes (table 6–1) ranging from 3/8-inch diameter to about 2 1/4-inch diameter, as wire for column spirals, and as wire mesh which is often used for temperature and shrinkage reinforcements of slabs and walls. Reinforcing steel is specified in ASTM A615, A616, and A617. The minimum yield strengths are: 40,000 psi, 50,000 psi, 60,000 psi, and 75,000 psi. The engineer will specify the size, amount and classification of reinforcing steel to be used. Wire mesh is cold-drawn from hard-grade steel. All steel used in the manufacture of concrete reinforcing bars is ductile. Generally, hooks of relatively small radius can be cold-formed on the job without breakage. Improved deformed bars, which conform to ASTM Specifications are so superior in bond value that hooking the ends of the bar does not add significantly to the strength. Bars which do not meet ASTM specifications are no longer rolled for concrete reinforcement. With so many grades of steel now available, quite likely having the same pattern of deformation, some permanent, rolled-on identification is necessary and is required by the latest ASTM specifications. Two systems of grade markings have been adopted and are illustrated in figure 6–7:

Continuous longitudinal smaller lines between the main longitudinal ribs. One small line identifies 60,000 psi; two such lines 75,000 psi yield strength.

Rolled-on numbers follow the symbols and number that indicates bar size. The number 40 identifies 40,000 psi; the number 50 identifies 50,000 psi; the number 60 identifies 60,000 psi, and the number 75 identifies 75,000 psi yield strength steel.

Hooks. It is not possible or feasible in some cases to extend straight bars far enough to develop their strength by bond only. In these cases, it is common practice to specify bent or hooked rods to obtain the required length for their development through bond. The hook provides some mechanical locking of the steel into the concrete. Good engineering judgment will frequently dictate means to insure suitable anchorage for the end of

*Table 6–1. Standard Steel Reinforcing Bars**

Bar designation No.**	Unit weight lb/ft.	Diameter in.	Cross-sectional area, in.^{2}	Perimeter in.
3	.376	0.375	0.11	1.178
4	.668	0.500	0.20	1.571
5	1.043	0.625	0.31	1.963
6	1.502	0.750	0.44	2.356
7	2.044	0.875	0.60	2.749
8	2.670	1.000	0.79	3.142
9	3.400	1.128	1.00	3.544
10	4.303	1.270	1.27	3.990
11	5.313	1.410	1.56	4.430
14	7.65	1.693	2.25	5.32
18	13.60	2.257	4.00	7.09

*The nominal dimensions of a deformed bar are equivalent to those of a plain round bar having the same weight per foot as the deformed bar.
**Bar numbers are based on the number of eighths of an inch included in the nominal diameter of the bars.

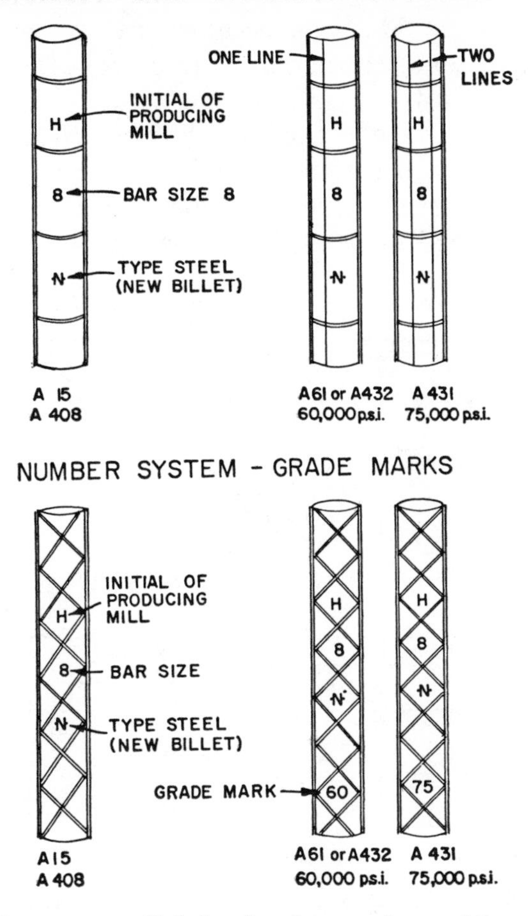

Figure 6–7. Reinforcing bar grade markings.

the rod, particularly if the bar has to resist large tension very near the end. Hooks are usually necessary for reinforcing bars in tension unless the bars terminate at the tops of continuous beams. When it is practicable, the longitudinal rods in a beam are anchored in the compression area. Details for bending hooks in reinforcing bars are shown in table 6–2. The 90° hook may be used when a little mechanical anchorage is desired but full development of the rod is not necessary. As indicated, a reasonably large radius is required for this type of hook. The 135° hook is less desirable than the 90° hook because the acute angle of the bend tends to increase the compression in the concrete at the inside of the bend. It is usually preferable to bend the rod 180° to withstand the compression caused by tension in the steel. The straight portion beyond the bend is also desirable as additional anchorage. All bends in reinforcing

Table 6–2. Hook Details for Reinforcing Steel

RECOMMENDED SIZES – 180° HOOK

d, J, EXT., D, 4d MIN, H

D = 6d for bars #2 to #7

D = 8d for bars #8 and #9

D = 10d for bars #10 and #11

BAR SIZE d	BAR EXTEN	J	APPROX. H
#3	5	3	4
4	6	4	4 1/2
5	7	5	5
6	8	6	6
7	10	7	7
8	13	10	9
9	15	11 1/4	10 1/4
10	19 1/2	15 1/4	12 3/4
11	21 1/2	17	14 1/4

MINIMUM SIZES – 180° HOOK

d, J, EXT., D, H

D = 5d MIN.

NOTE: MINIMUM SIZE HOOKS TO BE USED ONLY FOR SPECIAL CONDITIONS. DO NOT USE FOR HARD-GRADE STEEL.

BAR SIZE d	BAR EXTEN	J	APPROX. H
#3	5	2 3/4	4
4	5	3 1/2	4 1/4
5	6	4 1/4	4 3/4
6	7	5 1/4	5 3/4
7	9	6	6 1/2
8	10	7	7 1/2
9	11	8	8 1/2
10	13	9	9 1/2
11	14	10	10 1/2

RECOMMENDED MINIMUM SIZES – 90° HOOK

d, J, EXT., D, 12 d MIN.

D = 7d for #3 to #7

D = 8d for #8 and #9

D = 10d for #10, #11 #14S and #18S

BAR SIZE d	BAR EXTEN.	APPROX. J
#3	5 1/2	6
4	7 1/2	8 1/4
5	9	10 1/4
6	10 1/2	12 1/2
7	12 1/2	14 1/2
8	14 1/2	17
9	16 1/2	19
10	18 1/2	23
11	20 1/2	25 1/2
14S	25	30 1/2
18S	33	40 1/2

HOOKS FOR STIRRUPS AND TIES – 90° AND 135°

H, EXT., D

D = 1 1/2d

BAR SIZE d	90°		135°	
	HOOK A OR G	J	HOOK A OR G	APPROX. H
#3	3	3 1/4	4	2 1/2
4	3 1/2	4 1/4	4 1/2	3 1/4
5	4 1/2	5 1/2	5	3 3/4

NOTE: STIRRUP HOOKS MAY BE BENT TO THE DIAMETER OF THE SUPPORTING BARS.

bars should have reasonably large diameters so that the full length of the rod can be considered to be effective. A hook should not be used to anchor a bar subjected to compression.

Bolsters and Chairs. Bolsters or chairs (fig. 6–5) are available in a variety of heights to suit any situation. Bolsters or chairs keep the bars carrying computed stress from 1 to 2 inches from the outside shell. This is frequently called protective concrete.

Stirrups. Small U-shaped bars called stirrups are used to supplement bent bars in resisting diagonal tension and to reinforce the web in a beam to prevent cracks from spreading. They must pass underneath the bottom steel and be perpendicular to it to prevent lateral slippage. Vertical stirrups can be readily arranged and so easily set in the forms with the other rods that they are one of the most practical systems of web reinforcement. Anchorage of the stirrups is secured by welding to longitudinal steel, by hooking tightly around longitudinal reinforcement, and by embedding sufficiently above the mid-depth of the beam to develop the required stress by bond. Welded stirrups (fig. 6–4) may be placed at any angle. They serve the same purpose as vertical stirrups in resisting diagonal tension in the beam. Inclined stirrups are more efficient than vertical ones since they may be oriented parallel with diagonal stress. Some practical considerations offset this advantage. This type of stirrup should be welded to longitudinal reinforcement to avoid slippage and displacement during concrete placement, but this work is expensive and somewhat troublesome. Stirrups must be placed so that every potential diagonal tension crack is crossed by at least one stirrup.

Splices.

(1) *Methods.* The usual method of splicing reinforcing bars is by lapping the bars past each

other so that bond stress will transfer the load from one bar into the concrete and then into the other bar (fig. 6–8). The rods could be hooked but it is not always practicable or even desirable to bend them. The length of the lap is a matter of engineering judgment in considering the stresses anticipated in the beam, but approximates 24 to 36 rod diameters, depending on the size of the rod. For plain bars the minimum length of the lap shall be twice that for deformed bars. Splices should not be made at the points of maximum bending. It is usually best to locate splices beyond the center of the beam. When possible, splices should be staggered so that all the splices do not come at the same point. The method shown in 1, figure 6–8 is satisfactory when the spacing of the bars is large but it is undesirable in a beam or similar member having several closely spaced bars when the overlapped section sometimes interferes with proper encasement of rods and the filling of forms.

Types. The lap in a horizontal plane illustrated by 2, figure 6–8 is the most practical arrangement if the spacing provides enough clearance for the passage of aggregate. Both this method and the one shown in 3, figure 6–8 facilitate tying the bars to hold them in position during concreting. However, the tying does not add significantly to the strength of the splices. There is a possibility of air pockets and poor bond in the space under the junction between the bars. Lapping of rods in a vertical plane as shown in 3, figure 6–8 has the advantage of better encasement but the top rods do not fit the stirrups properly, and the beam has a smaller effective depth at one place than at another one. In practice, bars in this position are likely to be knocked down in the position shown in 2, figure 6–8. Reinforcing bars are not butted when one rod is not long enough for the span. *Except as shown on plans, no splicing of*

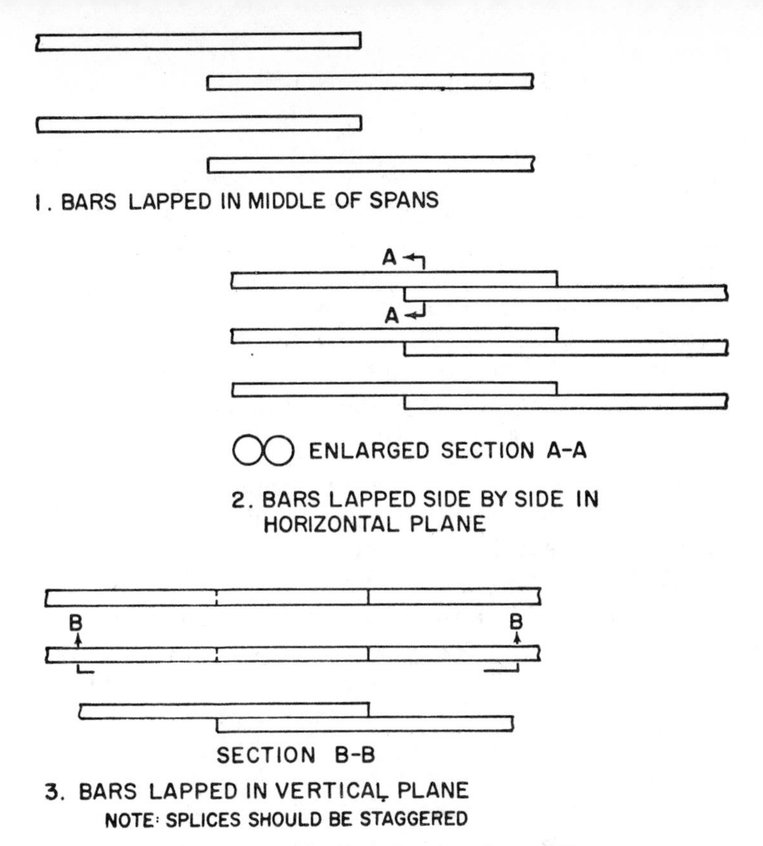

Figure 6–8. Reinforcing bar splices.

reinforcement should be made without approval of an engineer. There should be at least two supports under every bar and bolsters should be spaced at about 5-foot intervals.

Storing. Excessive rusting of the reinforcing steel in storage should be avoided. Before the steel is placed, the surface should be free from objectionable coatings, particularly heavy corrosion caused by outdoor storage. When stored outside, reinforcing bars should be placed on dunnage.

Cleaning. A thin film of rust or mill scale is not considered to be seriously objectionable—in fact, it may increase the bond of steel with concrete, but loose rust or scale which can be removed by rubbing with burlap or by other means should be removed. Other objectionable coatings commonly found on reinforcing steel are oil, paint, grease, dried mud, and weak dried mortar. If the mortar is difficult to remove, it will probably do no harm where it is. Anything that destroys the ability of the concrete to grip the steel may prove to be serious if it prevents the stress in the steel from performing its function properly.

Fabrication. When large numbers of reinforcing bars of varied lengths and shapes are required they can be prefabricated on the job according to the drawings. Stirrups and column ties are usually less than ½ inch in diameter and can be bent cold. Steel bars larger than ⅜ inch in diameter to be used for main reinforcement should be cold-bent. Heating is normally unnecessary except for bars over 1⅛ inch in diameter. The bends, except for hooks, should be made around pins having a diameter of not less than six times the bar diameter. If the bar is larger than 1 inch the minimum diameter of the bending pin should be eight times the bar size. Steel bars larger than

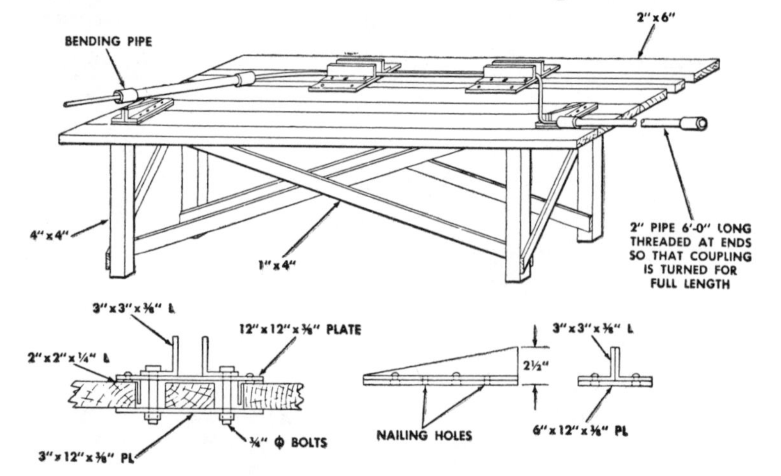

Figure 6–9. Bar bending table.

$3/8$ inch in diameter should be bent with a bar bending machine whenever possible. If a bending machine is not available, a hickey or the bar bending table shown in figure 6–9 may be used, although the bends made with these devices are usually too sharp and the bar is weakened. A hickey, a level for bending bars, may be improvised by attaching a 2- by $1\frac{1}{2}$- by 2-inch pipe tee to the end of a $1\frac{1}{4}$-inch pipe lever 3 feet long, and sawing a section from one side of the tee.

Placement

Support. All steel reinforcement should be accurately located in the forms and firmly held in place before and during the casting of concrete. This can be done by means of built-in concrete blocks, metallic supports, spacer bars, wires, or other devices adequate to insure against displacement during construction and to keep the steel at the proper distance from the forms. There should be enough supports and spacers to carry the steel properly even when subject to construction loads. The use of rocks, wood blocks or other unapproved objects to support the reinforcing steel is prohibited. Horizontal bars should be supported at minimum intervals of 5 or 6 feet. All bars should be secured to supports and to other bars by tie wires. Wire used to tie bars should not be smaller than 18 gage. The twisted ends of ties should project away from an interior surface.

Spacers. Some specifications require that no metal be left in the concrete within a given distance of the surface. Generally, the minimum clear distance between parallel bars in beams, footings, walls, and floor slabs should not be less than $1\frac{1}{3}$ times the largest size aggregate particle in the concrete nor less than 1 inch. Spacer and supporting blocks can be made from a mortar with the same consistency as the concrete, but without the

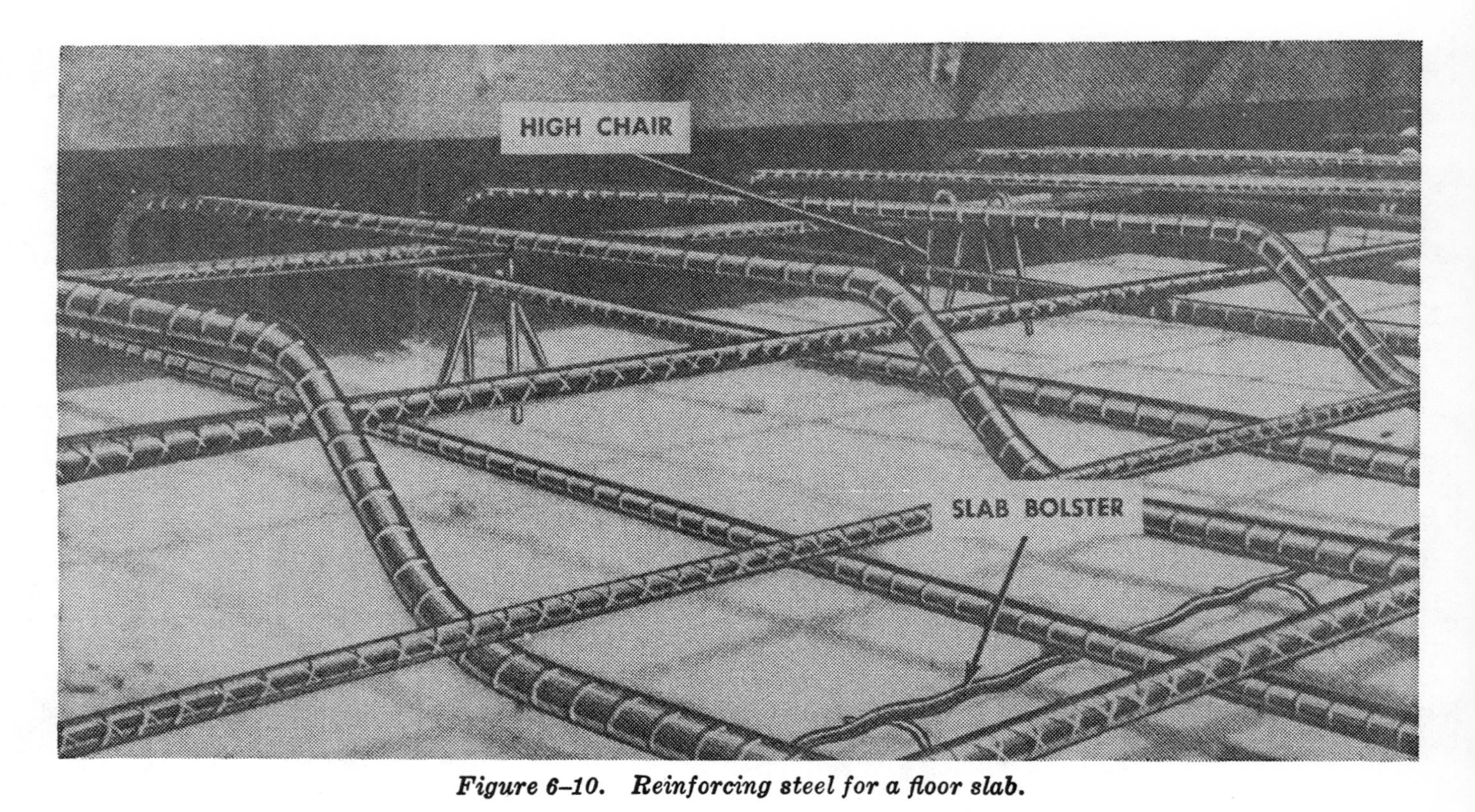

Figure 6–10. Reinforcing steel for a floor slab.

coarse aggregate. The spacer blocks are usually 1½ inches square or larger varying in length as required. Tie wires are cast in the blocks to secure the blocks to the reinforcing bars. When this type of spacer block is used, removal is not necessary when the concrete is placed.

Columns. The clear distance between parallel bars in columns should not be less than 1½ times the bar diameter. The steel for columns is first tied together and placed in position as a unit. Then the column form is erected around the unit and the reinforcing steel is tied to the form at 5-foot intervals.

Floor Slabs. A typical arrangement of reinforcing steel in floor slabs is shown in figure 6–10. The height of the slab bolster is determined by the thickness of the required concrete protective cover. Concrete blocks made of sand-cement mortar can be used in place of the slab bolster. The bars should be tied together with one turn of wire at frequent intervals where they cross, to hold them firmly in place.

Beams. Reinforcing steel for a reinforced concrete beam shown in figure 6–11, indicates the position of bolsters and stirrups. Note that the stirrups pass under the main reinforcing rods and are tied to them with one turn of wire.

Walls and Footings. Reinforcing steel is erected in place for walls, unless wire fabric is used. It is not preassembled as it is for columns. Ties between the top and bottom should be used for high walls. The wood blocks (1, fig. 6–12) are removed when the forms have been filled up to the level of the block. Welded wire fabric (2, fig. 6–12) is also used as concrete reinforcement for footing, walls and slabs. Reinforcing steel for footings should be placed after the forms have

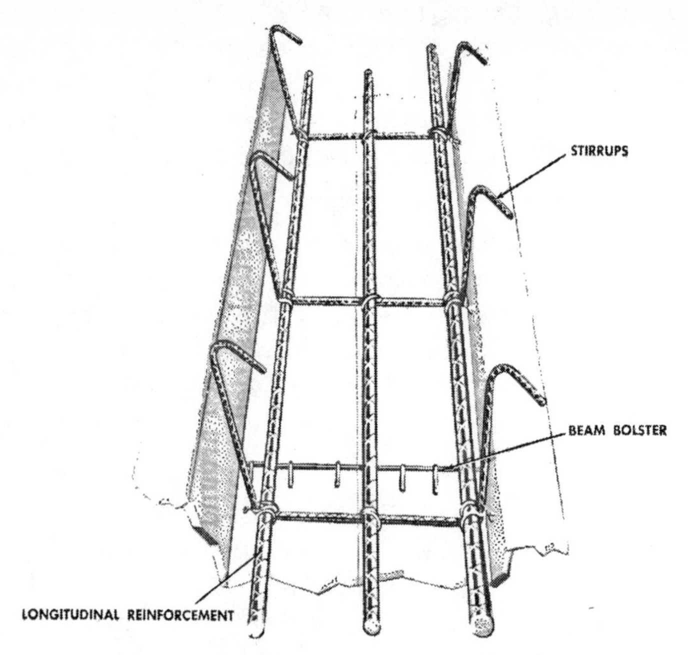

Figure 6–11. Beam reinforcing steel.

been set. A typical arrangement is shown in 3, figure 6–12. Concrete bars or clean sound stones may be used in footings to support the steel the proper distance above the subgrade.

PRECAST CONCRETE

Definition. Precast concrete is any concrete member that is cast in forms at a place other than its final position of use. The member may be of either plain or reinforced concrete. It can be done anywhere although this procedure is best adapted to a factory or yard. Job-site precasting is not uncommon for large projects. Some manufacturers produce a variety of structural members in several different shapes and sizes, including piles, girders, roof members and other standard products. Prestressed concrete is especially well adapted to precasting techniques.

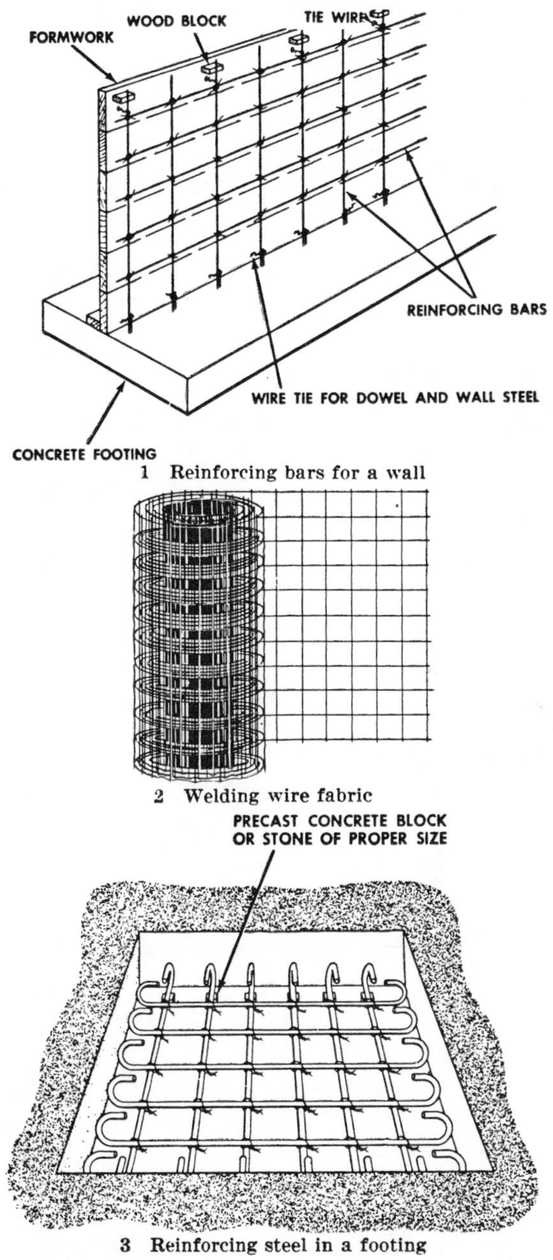

1 Reinforcing bars for a wall

2 Welding wire fabric

3 Reinforcing steel in a footing

Figure 6–12. Wall and footing reinforcement.

Products. Generally, structural members including standard highway girders, piles, electric poles, masts and building members are precast by factory methods unless the difficulty or impracticability of transportation makes job-site casting more desirable. The economies obtained by precasting standard members or members required in large numbers in a central location, for a particular project, are readily apparent.

Advantages. Economy of mass production is the principal advantage of precasting. Added to this is the desirability of fabricating on the ground rather than in the final position the member is intended for.

Disadvantages. Some of the inherent advantages of precasting are offset by the necessity for extensive plant facilities including equipment storage space. In addition, some advantages are offset by the cost of transporting and the necessity for heavy equipment to place precast members in position.

Transporting Precast Members. Prestressed members can be hauled further and given rougher treatment without detriment than either plain or reinforced members. In all cases, care must be

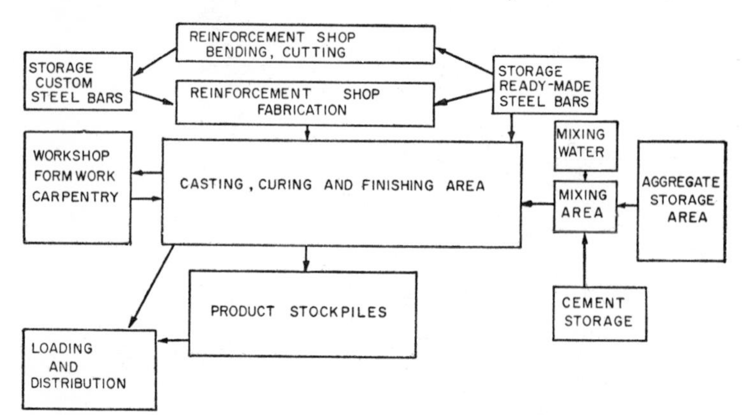

Figure 6–13. Schematic layout of prefabricated yard.

taken to support the members in such a way that no excessive strain or loading is applied that is different from the design loading. Various types of hauling and handling equipment are used in precast concrete operations. Heavy girders to be transported long distances can be hauled with tractor trailers or with a tractor and dolly arrangement in which the girder acts as the tongue or tie between the tractor and the dolly. Smaller members such as columns, piles, slabs, etc., can be hauled on flatbed trailers. These members must be protected from excessive bending stresses due to their own dead weight.

f. Erecting Precast Elements. The erection of precast members is similar to steel erection. Cranes or derricks of sufficient capacity are the usual means of lifting the members. Spreader bars frequently must be used in order to handle the elements at the correct pickup points. In some cases, due to long length or heavy weight, two cranes or derricks may be necessary to lift the members.

Prefabrication Yard

Precasting is done either in central prefabrication plants or at site prefabrication plants depending upon the product and its application. On site or temporary prefabrication plants are generally more suitable for military operations. These plants are without roofing and therefore subject to weather and climate considerations. The prefabrication yard is laid out to suit the type and quantity of members to be processed. It must be on firm level ground providing ample working space and access routes. Bridge T-beams, reinforced concrete arches, end walls, and concrete logs are typical members produced at these plants. A schematic layout of a prefabrication yard suitable for producing such members is shown in figure 6–13. The personnel requirements are given in table 6–3. A prefabrication unit of this size can be expected to produce approximately 6000 square feet of precast walls per day. The output will vary according to personnel experience, equipment capabilities, and product requirements.

Design

The design of a structure which specifies precast members requires the solution of problems inherent in this type of construction, including provisions for seating members and tieing the structure together that call for ingenuity and engineering skill. No attempt has been made to discuss the engineering or design of precast concrete since familiarization with the end products and the materials and methods used in producing them has been the purpose of this chapter.

Table 6–3. Recommended Precasting Team Composition

Area	Personnel
Renforcement shop and steel storage	1 Non Com 4 EM.
Workshop	2 EM.
Casting, curing, and finishing Placing reinforcement Refurbishing forms.	2 Non Com. 13 EM.
Mixing area Water, cement, and aggregate storage	1 Non Com. 5 EM.
Product stockpiles Loading and distribution.	2 EM.

PART TWO
MASONRY

7

General

MASONS' TOOLS AND EQUIPMENT

The mason's tools are shown in figure 7–1.

Trowels. The trowel is usually triangular, the largest size being 9 to 11 inches long and from 4 to 8 inches wide. The length and weight of the trowel used depends on the mason. He should select the one he can handle the best. Generally, the short wide trowels are best since the weight is nearer the wrist and does not put as much strain on the wrist. Trowels used for pointing and striking joints are smaller in size: 3 to 6 inches long and 2 to 3 inches wide. The trowel is used to—

Mix and pick up mortar from the board.

Throw mortar on the block.

Spread mortar.

Tap the block down into its bed when necessary.

Chisel or Bolster. The tool is used to cut concrete block. It is 2½ to 4½ inches wide.

Hammer. The hammer has a square face on one end and a long chisel peen on the other. It weighs from 1½ to 3½ pounds. It is used for splitting and rough-breaking of blocks.

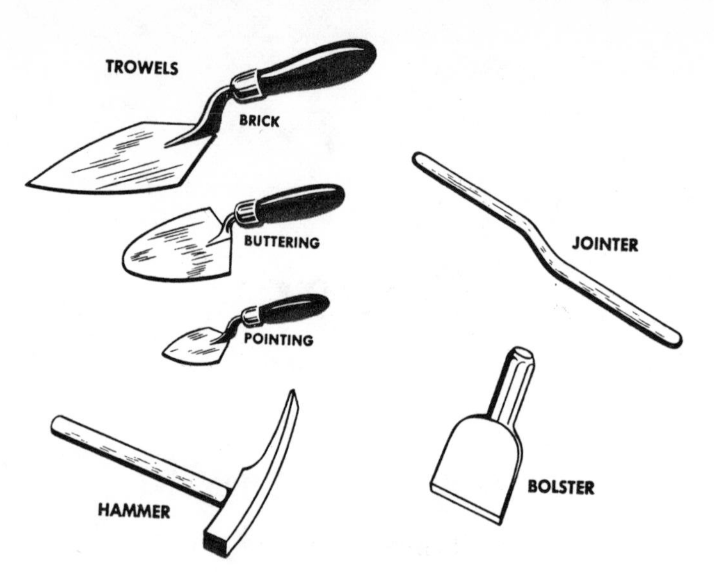

Figure 7–1. Mason's tools.

Jointer. This tool is used for making various types of joints. There are several different types. They are rounded, flat, or pointed depending on the shape of the mortar joint desired.

Square. The square (1, fig. 7–2) is used to measure right angles and lay out corners.

Mason's Level. The level enables the mason to plumb and level walls. It is from 36 to 48 inches long and is made of wood or metal. Two, figure 7–2 illustrates a level. When the level is placed horizontally on the masonry and the bubble in the center tube is exactly in the middle of the center tube, the masonry is level. When the level is placed vertically against the masonry and the bubble in the end tube is exactly in the middle of the tube, the masonry is plumb. For long, high walls or tall columns an offset line from the face of the work should be established. To assure straightness and plumbness, offset checks between this line and the face should be made frequently.

Straightedge. A straightedge, as shown in 3, figure 7–2, may be of any length up to 16 feet and should be 1⅛ inches thick and 6 to 10 inches wide. The top and bottom edges must be parallel. The straightedge can be used as an extension of the level to cover distances longer than the length of the level.

Miscellaneous Tools. Additional equipment required includes shovels, mortar hoes, wheelbarrows, chalk, plumb bobs, and a 200-foot ball of No. 18 to 21 hard-twisted cotton cord of the same type used as chalkline by carpenters.

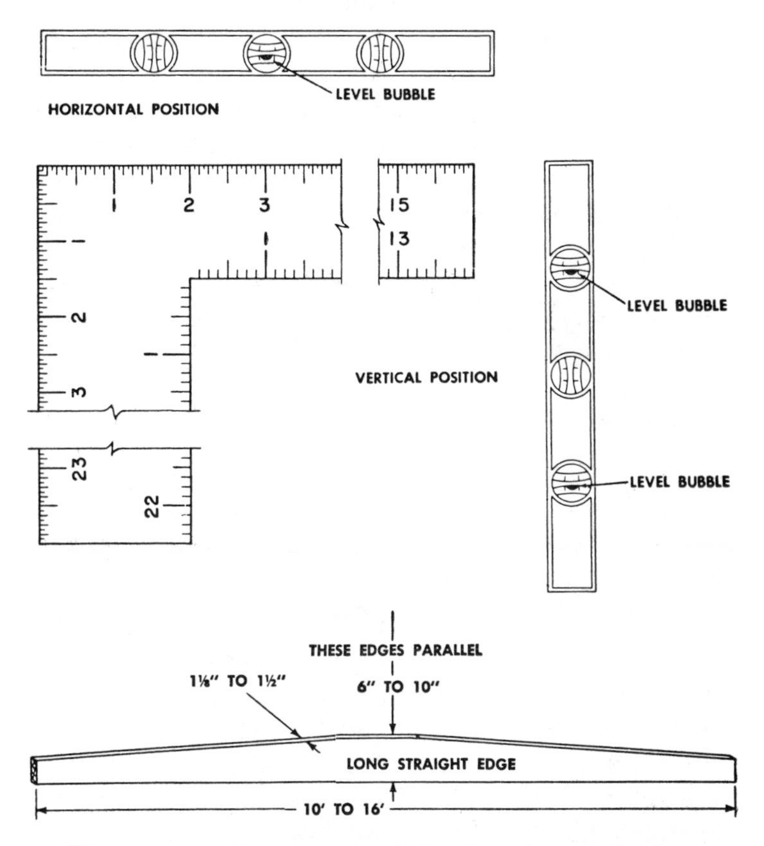

Figure 7–2. Square, mason's level, and straightedge.

Care of Tools. Wheelbarrows, mortar box, and mortar tools should be kept clean since hardened mortar is difficult to remove. All tools and equipment should be cleaned thoroughly at the end of each day, or when the job is finished, as appropriate.

Equipment

Mortar Box. The mortar box (1, fig. 7–3) is used to mix mortar by hand. It should be as water tight as possible.

Mortar Board. The mortar board is constructed as shown in 2, figure 2–3. It can be from 3- to 4-feet square. The mortar board should be thoroughly wetted down before any mortar is placed on it to prevent the wood from absorbing moisture and causing the mortar to dry out. The mortar should be kept rounded up in the center of the board and the outer edges kept clean. If spread in a thin layer, the mortar will dry out quickly and there will be a tendency for lumps to form. Proper consistency must be maintained at all times. Three, figure 7–3 indicates the proper way to fill a mortar board.

MORTAR

Basic Considerations. Good mortar which is necessary to good workmanship and good wall service must bond the masonry units into a strong, well-knit wall. The strength of the bond is affected by various factors including the type and quantity of the cementing material, the workability or plasticity of the mortar, the surface texture of the mortar bedding areas, the water retentivity of the mortar, and the quality of workmanship in laying the units. Water retentivity is that property of mortar which resists rapid loss of water to masonry units which may possess high absorption.

Mortar which is used to bond brick together will be the weakest part of brick masonry unless properly mixed and applied. Both the strength and resistance to rain penetration of brick masonry walls are dependent to a great degree on the strength of the bond. Water in the mortar is essential to the development of bond and if the mortar contains insufficient water the bond will be weak and spotty. When brick walls leak it is usually through the mortar joints. Irregularities in dimensions and shape of bricks are corrected by the mortar joint.

Mortar Properties. Mortar should be plastic enough to work with a trowel. The properties of mortar depend largely upon the type of sand used in it. Clean, sharp sand produces excellent mortar. Too much sand in mortar will cause it to segregate, drop off the trowel, and weather poorly. Workability, an important requirement of mortar, should be obtained through the proper grading of the sand, the use of mortar with good water retentivity, and through thorough mixing rather than through the use of excessive amounts of cementitious material.

Water Retentivity. Loss of moisture due to poor water retention results in rapid loss of plasticity and may seriously reduce the effectiveness of the bond. As concrete masonry units should be kept dry until they are built into the wall, they should never be wetted to control suction before the application of mortar.

Strength and Durability. The strength and durability requirements of a mortar depend upon the type of service the wall is to give. Walls subjected to severe stresses or to severe weathering naturally need to be laid in more durable, stronger

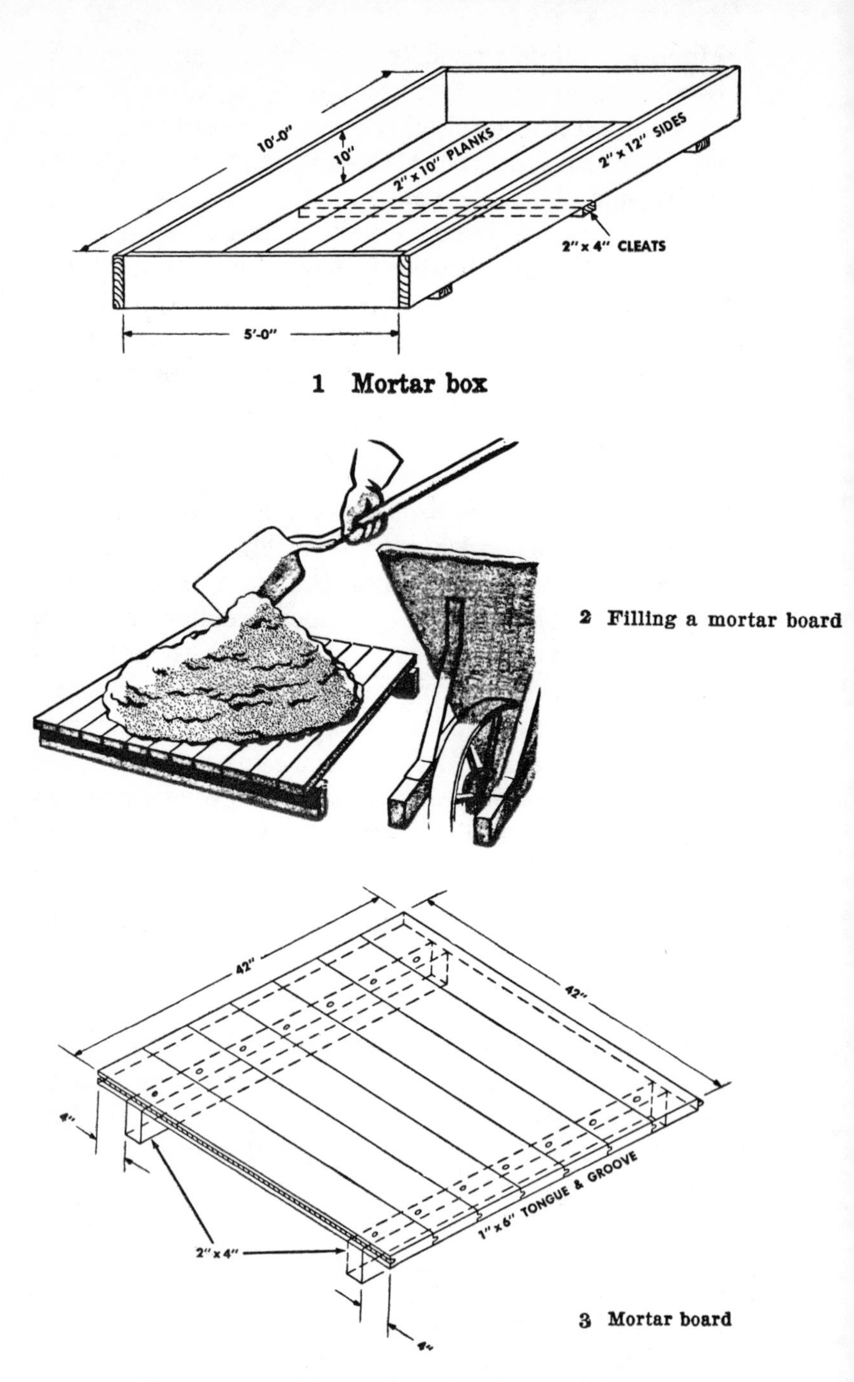

Figure 7–3. Mortar board and mortar box.

mortars than walls for ordinary service. Table 7–1 lists mortar mixes that provide adequate mortar strength and durability for the conditions indicated. The volumetric proportions shown may be converted to weight proportions by multiplying the unit volumes by the weight per cubic foot of the materials, which may be assumed to be as follows:

Masonry cement	Weight printed on bag
Portland cement	94 lb
Hydrated lime	40 lb
Mortar sand, damp and loose	85 lb (approximately)

The section of mortar for brick construction depends on the use requirements of the structure. For example, the recommended mortar for use in laying up interior non-load-bearing partitions would not be satisfactory for foundation walls. In many cases, the builder relies upon a fixed proportion of cement, lime and sand to provide a satisfactory mortar.

Bond. Bond is the property of a hardened mortar that knits the masonry units together. The strength of bond is affected by a number of factors such as the kind and quantity of cementitious material, the workability of the mortar, the surface texture of the mortar bedding areas, the rate of suction of the masonry units and the quality of workmanship in making the joints.

Types of Mortar. The following types of mortar are proportioned on a volume basis:

Type M. 1 part portland cement, 1/4 part hydrated lime or lime putty, 3 parts sand, or 1 part portland cement, 1 part type II masonry cement, and 6 parts sand. This mortar is suitable for general use and is recommended specifically for masonry below grade and in contact with earth, such as foundations, retaining walls, and walks.

Type S. 1 part portland cement, ½ part hydrated lime or lime putty, 4½ parts sand, or ½ part portland cement, 1 part type II masonry cement and 4½ parts sand. This mortar is also suitable for general use and is recommended where high resistance to lateral forces is required.

Table 7-1. Recommended Mortar Mixes
Proportions by volume.

Type of service	Cement	Hydrated lime	Mortar sand, in damp, loose condition
FOR ORDINARY SERVICE.	1—masonry cement*	----------	2¼ to 3.
	or 1—portland cement	½ to 1¼	4½ to 6.
SUBJECT TO EXTREMELY HEAVY LOADS, VIOLENT WINDS, EARTHQUAKES, OR SEVERE FROST ACTION. ISOLATED PIERS.	1—masonry cement* plus 1—portland cement	----------	4½ to 6.
	or 1—portland cement.	0 to ¼	2¼ to 3.

*ASTM Specification C 91 Type II.

Type N. 1 part portland cement, 1 part hydrated lime or lime putty, 6 parts sand, or 1 part type II masonry cement and 3 parts sand. This mortar is suitable for general use in exposed masonry above grade and is recommended specifically for exterior walls subjected to severe exposures as, for example, on the Atlantic Seaboard.

Type O. 1 part portland cement, 2 parts hydrated lime or lime putty, and 9 parts sand, or 1 part type I or type II masonry cement and 3 parts sand. This mortar is recommended for load-bearing walls of solid units where the compressive stresses do not exceed 100 pounds per square inch and the masonry will not be subjected to freezing and thawing in the presence of excessive moisture.

Storage of Mortar Materials. All mortar materials except sand and slaked quicklime must be stored in a dry place.

Mixing Mortar

Machine Mixing. If a large quantity of mortar is required, it should be mixed in a drum-type mixer similar to those used for mixing concrete. The mixing time should not be less than 3 minutes. All dry ingredients should be placed in the mixer first and mixed for 1 minute before adding the water.

Hand Mixing. Unless large amounts of mortar are required, the mortar is mixed by hand using a mortar box shown in figure 7–3. Care must be taken to mix all the ingredients thoroughly to obtain a uniform mixture. As in machine mixing, all dry material should be mixed first. A steel drum full of water should be kept close to the mortar box for the water supply. A second drum of water should be available for shovels and hoes when not in use.

Mixing Mortar With Lime Putty. When a machine mixer is used, the lime putty should be measured with a GI pail and loaded in the skip on top of the sand. If the mortar is to be mixed by hand, sand is added to the lime putty. Pails should be wet before mortar is placed in them and should be cleaned immediately after they have been emptied.

Water. Water used in mixing mortar should meet the same requirements as water used in mixing concrete. Water containing large amounts of dissolved salts should not be used as they will cause efflorescence and weaken the mortar.

Retempering Mortar. Mortar that has stiffened on the mortar board because of evaporation should be reworked to restore its workability by thorough remixing and by the addition of water as required. Mortar stiffened by initial setting should be discarded. A practical guide in determining the suitability of mortar, since it is difficult to determine the cause of stiffening, is that it should be used within 2½ hours after original mixing when the air temperature is 80°F. or higher and within 3½ hours when the air temperature is below 80°F. Mortar not used within the above limits should be discarded.

Antifreeze Materials. The use of an admixture to lower the freezing point of mortars during winter construction should be avoided. The quantity of such material necessary to lower the freezing point of mortar to any appreciable degree would be so large that the mortar strength and other desirable properties would be seriously impaired. If mortar freezes, it must not be used; freezing destroys its bonding ability.

Accelerators. Calcium chloride is sometimes added to mortar to accelerate the rate of hardening and to increase early strengths. Not more than 2 percent calcium chloride by weight of the portland cement should be used for this purpose. Not more than 1 percent of calcium chloride should be used with masonry cements. A trial mix will indicate the percentage of calcium chloride that will give the desired rate of hardening of the mortar. High early strengths in mortars can also be obtained by the use of high-early-strength portland cement.

Repair and Tuckpointing. The mortar mixes shown in table 7–1 can be used in repairing and tuckpointing old masonry walls. After the mortar has partially stiffened, the joints should be thoroughly compacted by tooling.

SCAFFOLDING

Basic Considerations

A scaffold is a temporary platform built for the support of workman and materials. Scaffolds are necessary after the bricklayer has completed work at the height he can reach by standing on the floor or ground. Extreme care is taken in building scaffolds because workmen's lives depend upon them. No scaffolding is temporarily nailed for it may be forgotten and never adequately nailed. When the scaffold planks at one level are no longer needed, they should be removed; falling mortar will hit them and splash on the wall. Rough lumber should be used for wood scaffolding.

Types of Scaffolding

Several types of scaffolds are described below.

Trestle Scaffold. When construction is such that the brick can be laid from the inside of the wall, a trestle scaffold as shown in figure 7–4 may be used. The trestles should be from 4 to 4 feet 6 inches high. The scaffold planks rest on the trestles and should be 2 by 10's. After the wall has been built to a height of 4 or 5 feet, the trestle scaffold should be erected. The wall can then be completed to the next floor level while the bricklayer works from the scaffold. As soon as the rough flooring for the next floor is in place, the above procedure is repeated. The trestle should remain at least 3 inches from the wall in order to make sure it will not push against the newly laid brick and force them out of line.

Foot Scaffold. At times it may be necessary to reach higher than the trestle scaffold permits. Then a foot scaffold such as the one shown in figure 7–5 can be used. The 2 by 10 planks rest on bricks which can be supported by the trestle scaffold. This type of scaffold should not exceed 18 inches in height.

Putlog Scaffold.

When it is necessary to erect the scaffold from the ground to the height required, a putlog scaffold can be used to advantage. The uprights should be 4 by 4's supported on a 2- by 12-inch plank 12 inches long for bearing on the soil. These uprights should be spaced on 8-foot centers. There should be 4 feet 6 inches between the wall and the uprights. The ledgers should be made from 1- by 8-inch lumber nailed to the uprights, as shown in figure 7–6. The putlog is a 3- by 4-inch piece of lumber that rests on top of the ledger and against the upright. The other end of the 3 by 4 rests on the wall; a brick is omitted to provide an opening for it. The putlog is not fastened to the ledger. On top of the putlog, five 2 by 12's are placed to form the scaffold platform. The planks are not nailed to the putlog.

The uprights must be tied to the wall by stays. These stays may be passed through a window opening and fastened to the structure inside the building or spring stays may be used as shown in figure 7–6. Spring stays are made by placing two 2- by 6-inch boards in an opening in the wall formed by omitting a brick. After the boards are inserted into the hole, a brick is placed between them and forced to a position close to the wall. The boards are then sprung together and securely nailed to the ledger.

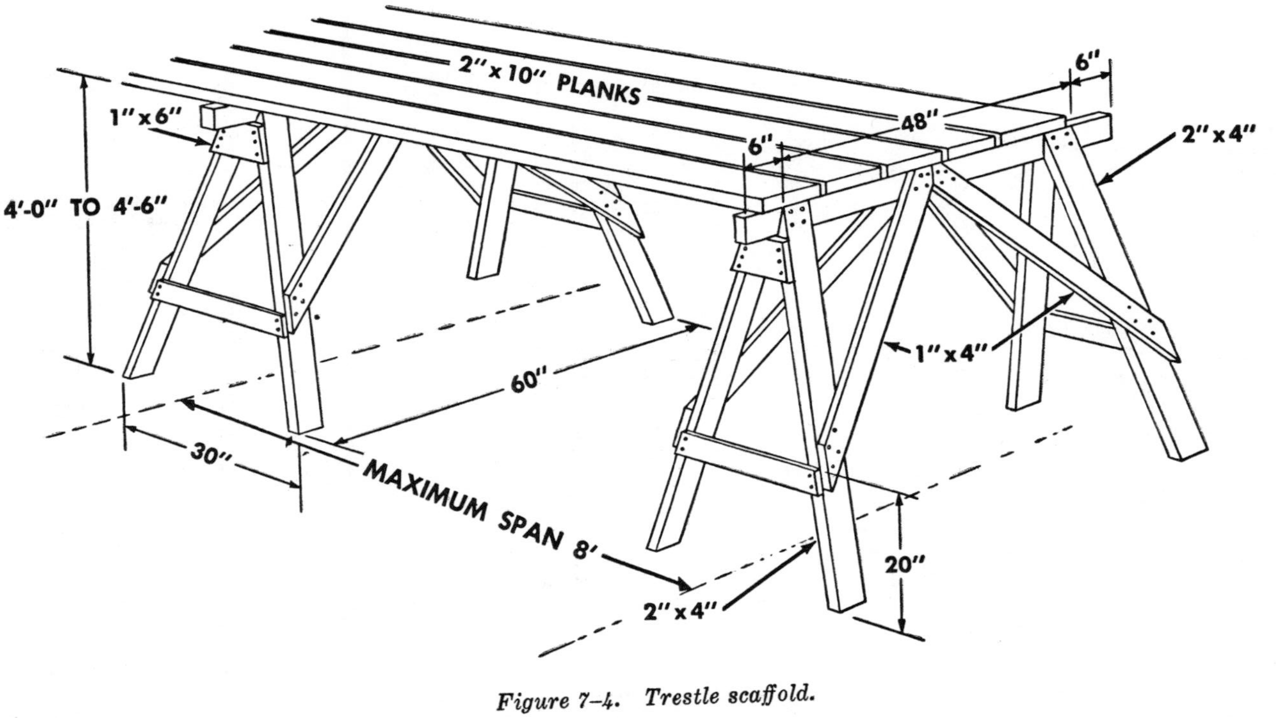

Figure 7–4. Trestle scaffold.

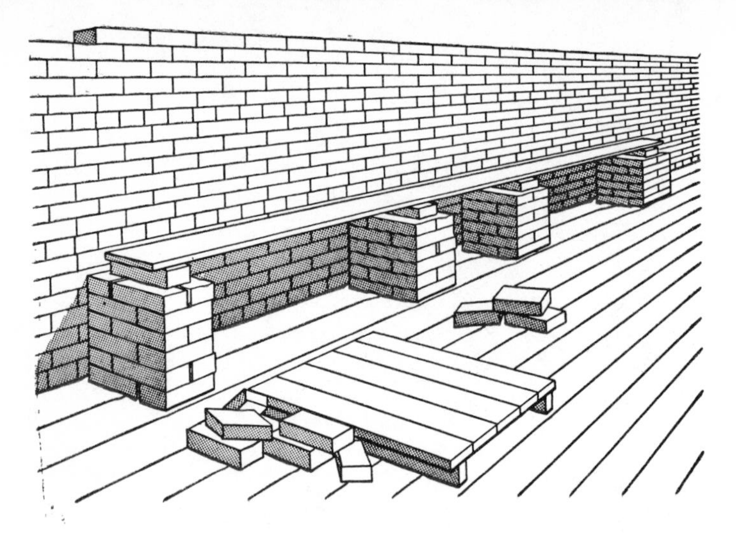

Figure 7–5. Foot scaffold.

The putlog may also be used as a stay in which case a wooden wedge should be driven above the putlog and into its hole in the wall. The wedge should then be nailed to the putlog and the putlog nailed to the ledger. Longitudinal cross bracing must be installed as shown in figure 7–8.

Outrigger Scaffold. This scaffold consists of 2- by 10-inch planks supported on a wooden beam projecting from the building. The beam is supported as shown in figure 7–7. If a steel outrigger beam is used, the beam is fastened to the form work of the structure by means of threaded U-shaped bolts.

Material Hoist

Details. On a large number of jobs, a material hoist is necessary. This hoist should be constructed as shown in figure 7–8. It should be located in such a way that materials may be moved to it with the shortest possible haul. The material tower should be located far enough away from the structure to clear any outside scaffold to

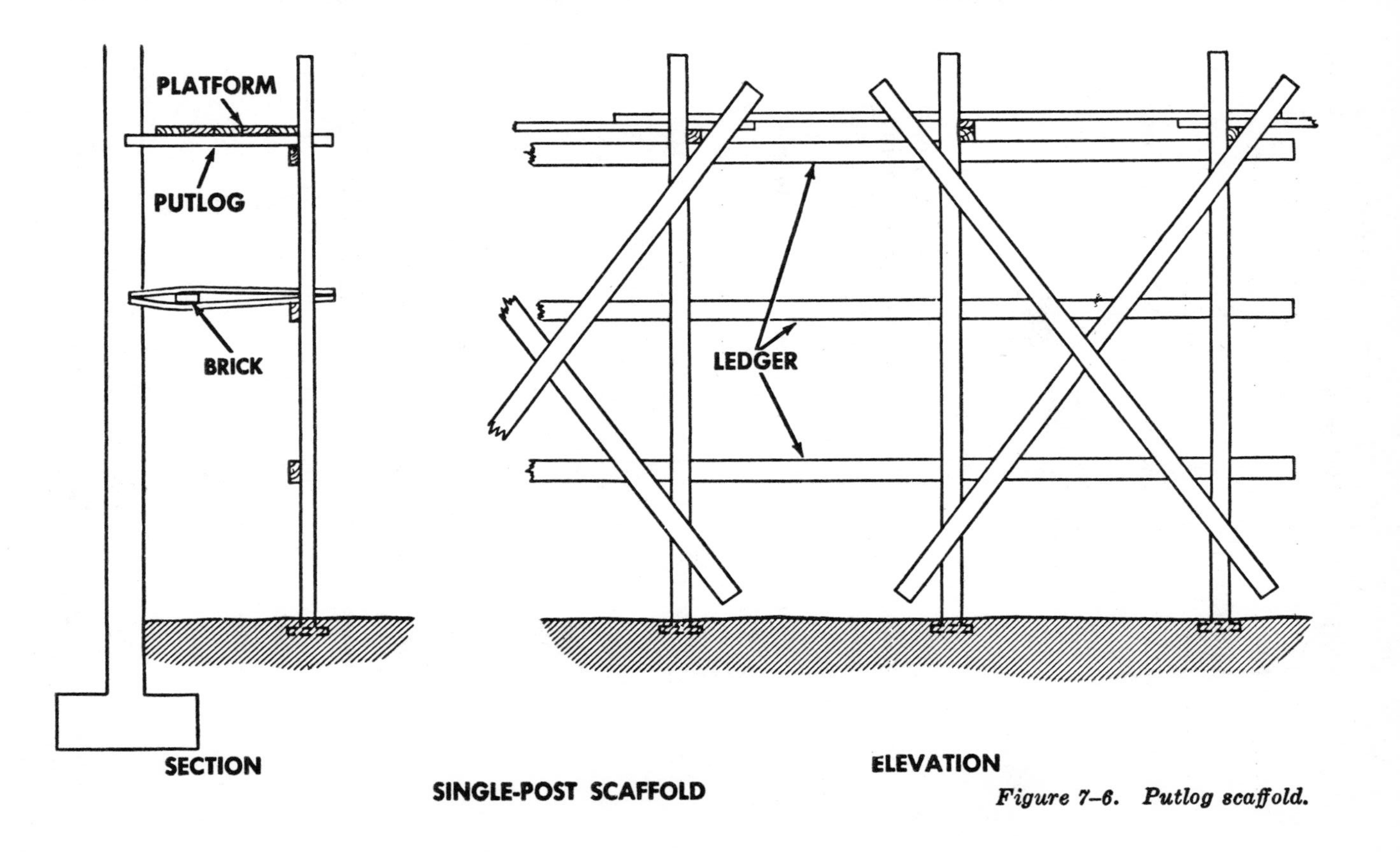

Figure 7–6. Putlog scaffold.

be used. A distance of 6 feet 8 inches is enough for scaffolds that have a platform made up of five 2 by 12's. Landings that extend from the material tower to the floors and scaffold platforms are constructed as needed. These landings should be made using 2- by 10- or 2- by 12-inch lumber. The tower should extend to a height of at least 15 feet above the highest point at which a landing is needed. The footing for the tower should be constructed of two 2 by 12's, 2 feet long, placed under each of the posts.

Elevator. The elevator for the material tower is shown in figure 7–9. Note how the elevator fits into the guides fastened to the material tower. The rope and pulley arrangement to be used is given in figure 7–9. If a steel tower is available, it should be used. Steel towers are more easily erected and generally safer.

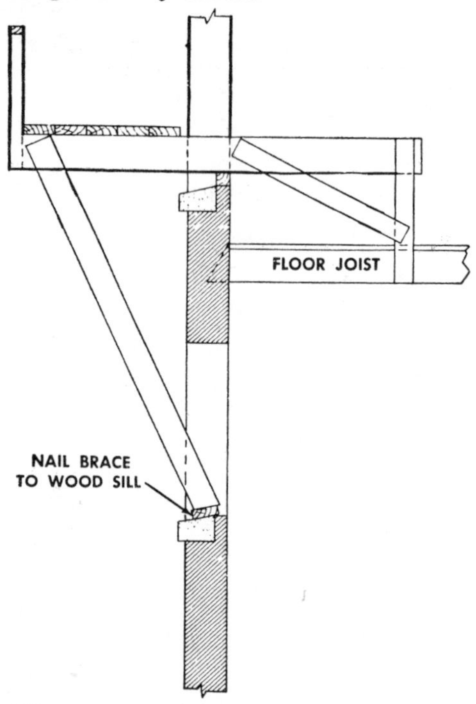

Figure 7–7. Outrigger scaffold.

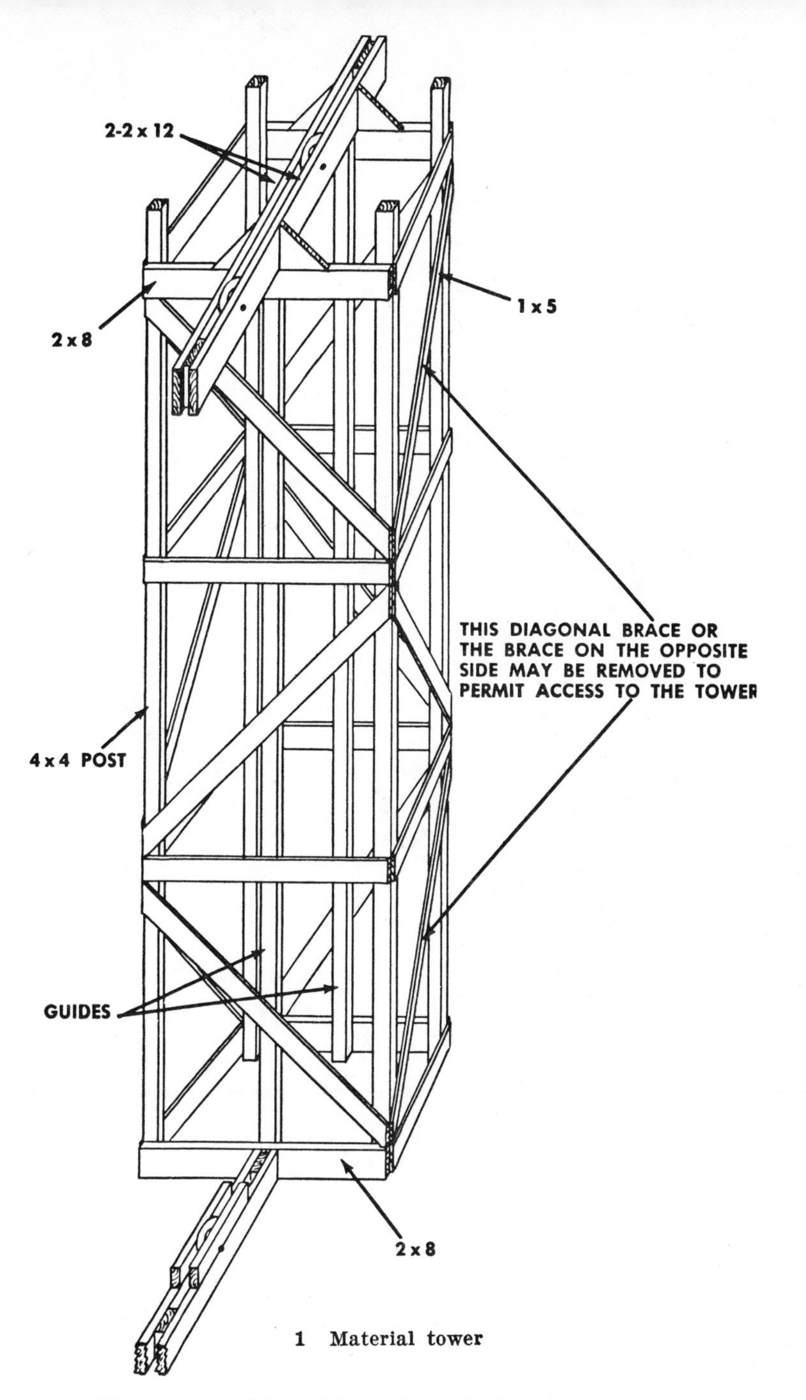

1 Material tower

Figure 7–8. Material tower and elevator.

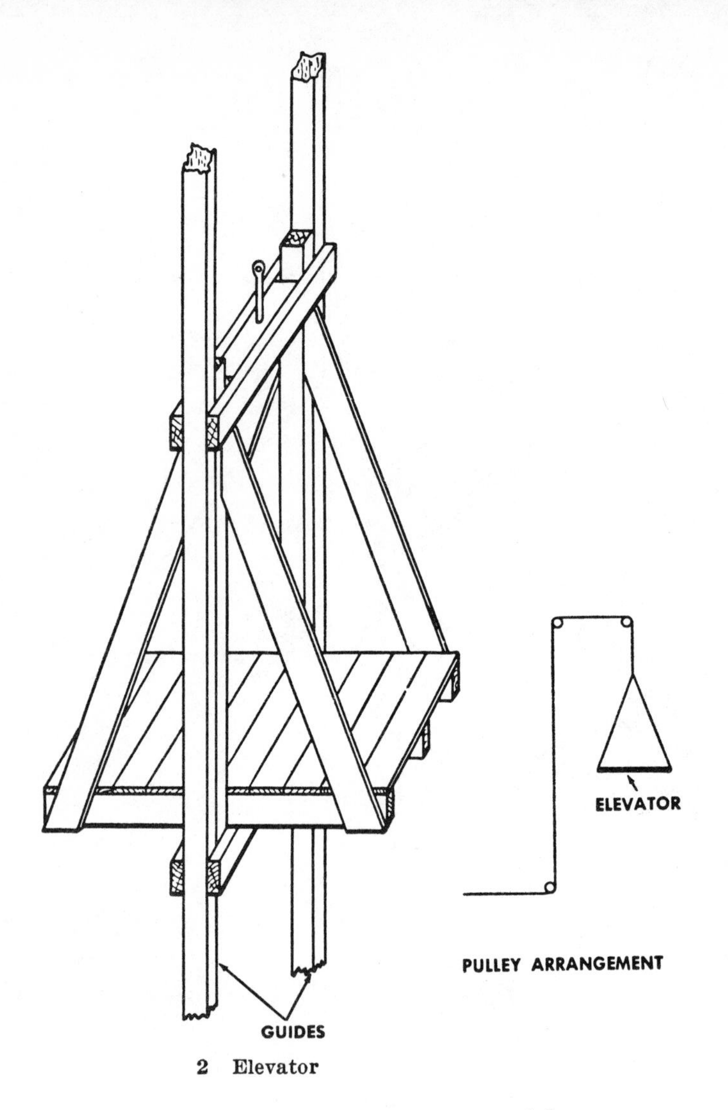

2 Elevator

Figure 7–9. Elevator for material tower

8

Concrete Masonry

CHARACTERISTICS OF CONCRETE BLOCKS

Concrete masonry has become increasingly important as a construction material. Important technological developments in the manufacture and utilization of the units have accompanied the rapid increase in the use of concrete masonry. Concrete masonry walls properly designed and constructed will satisfy varied building requirements including fire, safety, durability, economy, appearance, utility, comfort, and good acoustics.

Concrete Masonry Units

Uses. Concrete masonry units are designed and made for use in all types of masonry construction. Some of the uses are:

Exterior load-bearing walls (below and above grade).

Interior load-bearing walls.

Fire walls, curtain walls.

Partition and panel walls.

Backing for brick, stone, and other facings.

Fireproofing over structural members.

Firesafe walls around stair wells, elevators and other enclosures.

Piers and columns.

Retaining walls.

Chimneys.

Concrete floor units.

Types of Units. Concrete masonry building units are designated as:

Hollow load-bearing concrete block.

Solid load-bearing concrete block.

Hollow non-load-bearing concrete block.

Concrete building tile.

Concrete brick.

Heavyweight and Lightweight Units. The different types of units are made with heavyweight or lightweight aggregates and are referred to as heavyweight and lightweight units respectively. A hollow load-bearing concrete block of 8 × 8× 16 inches nominal size will weigh from 40 to 50 pounds when made with heavy weight aggregate such as sand, gravel, crushed stone or air-cooled slag. Concrete blocks made with lightweight aggregate will weigh from 25 to 35 pounds each and are made with coal cinders, expanded shale, clay, slag, or natural lightweight materials such as volcanic cinders and pumice. Heavyweight and lightweight units are used for all types of masonry construction. The choice of units depends on availability and the requirements of the structure under consideration.

Solid and Hollow Units. A solid concrete block is defined in ASTM specifications as a unit in which the core area is not more than 25 percent of the gross cross-sectional area. Concrete blocks are generally solid and are sometimes available with a recessed pocket called a "frog". A hollow concrete block is a unit having a core area greater than 25 percent of its gross cross-sectional area.

Generally, the core area of hollow units is 40 to 50 percent of the gross area.

Sizes and Shapes. Concrete building units are made in sizes and shapes to fit different construction needs. Units are made in full and half-length sizes as shown in figure 8–1. Concrete unit sizes are usually referred to by their nominal dimensions. A unit measuring 7⅝ inches wide, 7⅝ inches high and 15⅝ long is referred to as an 8 × 8 × 16 inch unit. When it is laid in a wall with ⅜-inch mortar joints, the unit will occupy a space exactly 16 inches long and 8 inches high. Local manufactures should be contacted for a schedule of sizes and shapes that are available. This information should be known prior to designing the proposed structure.

Block Machines. Blocks are usually made in a power-tamping machine. Machines of this type are available from several manufacturers. The concrete is tamped into a mold and the mold immediately stripped off. In this way, blocks can be rapidly made using only one mold. The mix used is dry enough so that the block will retain its shape.

Hand Method. In this method, concrete of fluid consistency is placed into sets of iron molds. The molds are stripped after the concrete has hardened. By this process, dense block can be made with little labor. The disadvantage lies in the fact that a large number of molds is required.

Curing. Concrete blocks are usually steam-cured since less time is required. Concrete blocks cured in wet steam at 125° F. for a period of 15 hours will have 70 percent of their 28-day strength. If steam is not available, blocks may be cured by protecting them from the sun and keeping them damp for a period of 7 days.

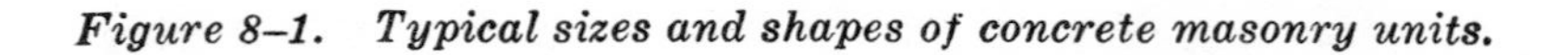

Figure 8–1. Typical sizes and shapes of concrete masonry units.

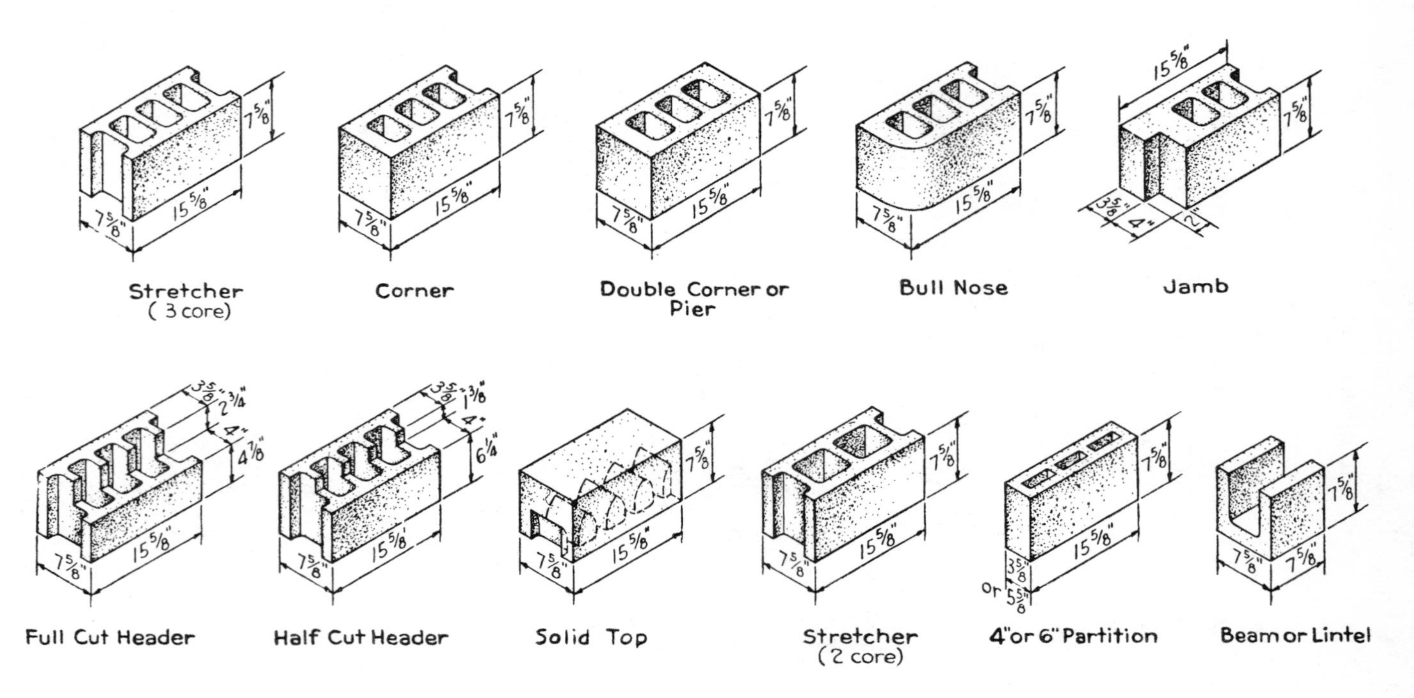

(Dimensions shown are actual unit sizes. A 7⅝″ x 7⅝″ x 15⅝″ unit is commonly known as an 8″ x 8″ x 16″ block.)

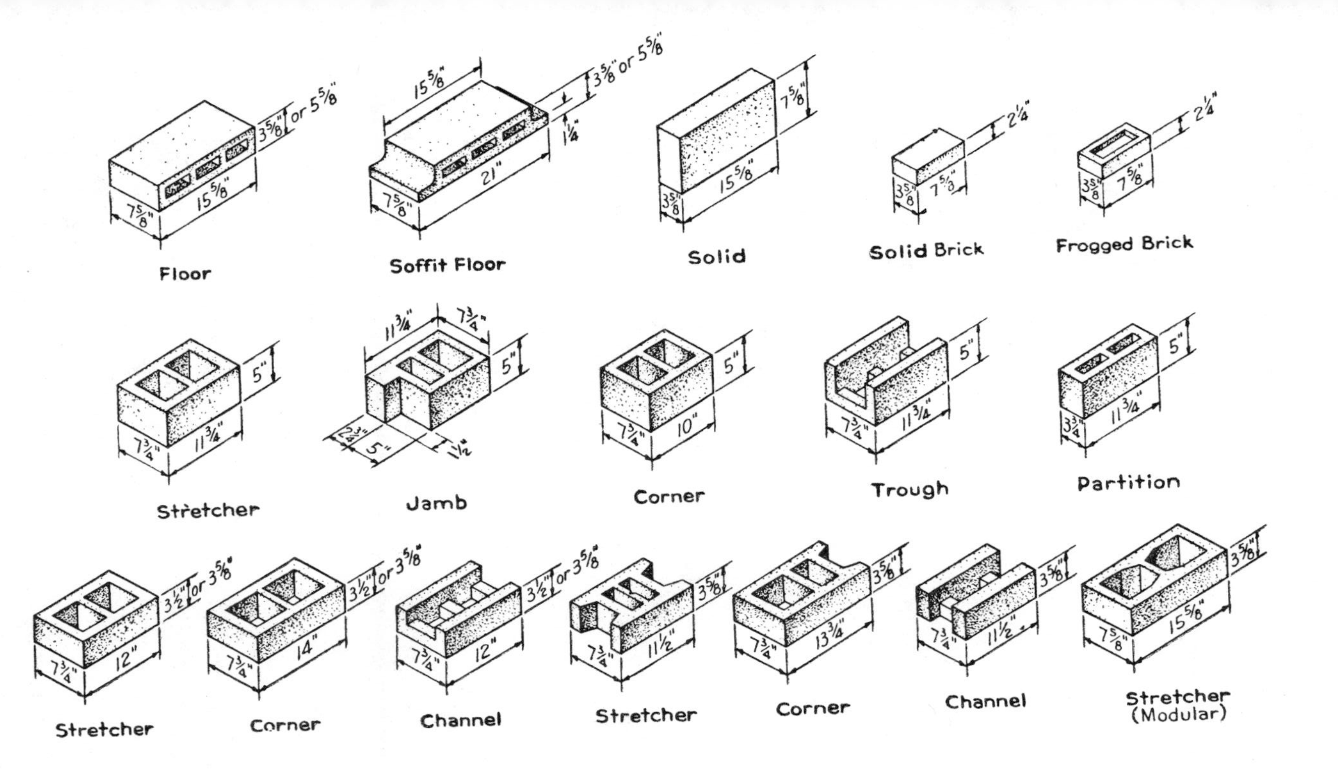

Floor
Soffit Floor
Solid
Solid Brick
Frogged Brick
Stretcher
Jamb
Corner
Trough
Partition
Stretcher
Corner
Channel
Stretcher
Corner
Channel
Stretcher (Modular)

i. Exposed Blocks. Concrete blocks exposed to weathering should be made with concrete having at least six sacks of cement per cubic yard of concrete. When lightweight porous aggregate is used, premixing with water for 2 minutes before adding the cement is advisable.

CONSTRUCTION PROCEDURES

Modular Planning

Concrete masonry walls should be laid out to make maximum use of full- and half-length units, thus minimizing cutting and fitting of units on the job. Length and height of wall, width and height of openings and wall areas between doors, windows, and corners should be planned to use full-size and half-size units which are usually available (fig. 8–2). This procedure assumes that window and door frames are of modular dimensions which fit modular full- and half-size units. Then, all horizontal dimensions should be in multiples of nominal full length masonry units and both horizontal and vertical dimensions should be designed to be in multiples of 8 inches. Table 8–1 lists nominal length of concrete masonry walls by stretchers and table 8–2 lists nominal height of concrete masonry walls by courses. When units $8 \times 4 \times 16$ are used, the horizontal dimensions should be planned in multiples of 8 inches (half-length units) and the vertical dimensions in multiples of 4 inches. If the thickness of the wall is greater or less than the length of a half unit, a special length unit is required at each corner in each course.

Footings

Masonry wall footings should be placed on firm, undisturbed soil of adequate load-bearing capacity to carry the design load and they should be below frost penetration. Unless local requirements or codes stipulate otherwise, it is general practice to make footings for small buildings twice as wide as

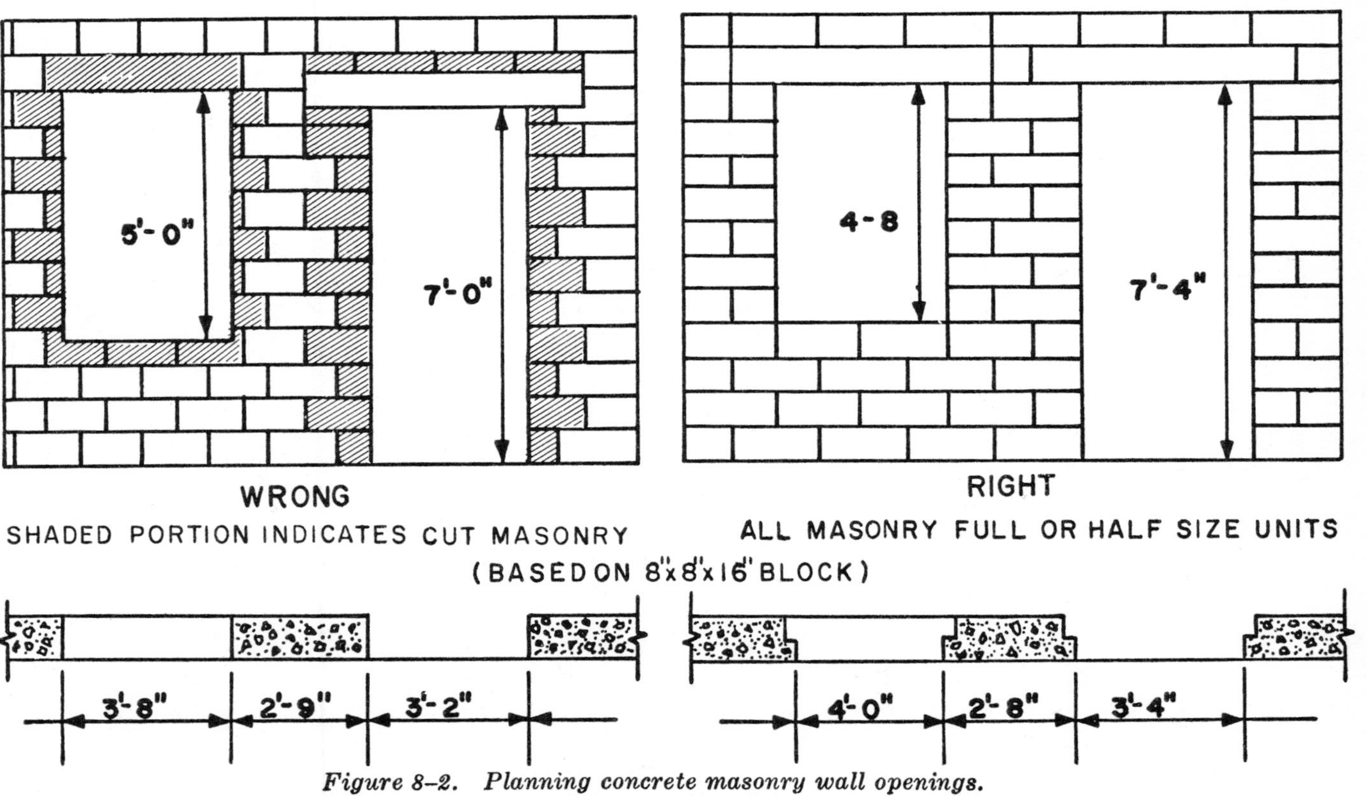

Figure 8–2. Planning concrete masonry wall openings.

the thickness of the walls they support. Table 8–3 lists weights and quantities of materials for concrete masonry walls. The thickness of the footings is equal to one-half their width (fig. 8–3).

Table 8–1. Nominal Length of Concrete Masonry Walls by Stretchers

No. of stretchers	Nominal length of concrete masonry walls	
	Units 15⅝″ long and half units 7⅝″ long with ⅜″ thick head joints.	Units 11⅝″ long and half units 5⅝″ long with ⅜″ thick head joints.
1	1′ 4″	1′ 0″.
1½	2′ 0″	1′ 6″.
2	2′ 8″	2′ 0″.
2½	3′ 4″	2′ 6″.
3	4′ 0″	3′ 0″.
3½	4′ 8″	3′ 6″.
4	5′ 4″	4′ 0″.
4½	6′ 0″	4′ 6″.
5	6′ 8″	5′ 0″.
5½	7′ 4″	5′ 6″.
6	8′ 0″	6′ 0″.
6½	8′ 8″	6′ 6″.
7	9′ 4″	7′ 0″.
7½	10′ 0″	7′ 6″.
8	10′ 8″	8′ 0″.
8½	11′ 4″	8′ 6″.
9	12′ 0″	9′ 0″.
9½	12′ 8″	9′ 6″.
10	13′ 4″	10′ 0″.
10½	14′ 0″	10′ 6″.
11	14′ 8″	11′ 0″.
11½	15′ 4″	11′ 6″.
12	16′ 0″	12′ 0″.
12½	16′ 8″	12′ 6″.
13	17′ 4″	13′ 0″.
13½	18′ 0″	13′ 6″.
14	18′ 8″	14′ 0″.
14½	19′ 4″	14′ 6″.
15	20′ 0″	15′ 0″.
20	26′ 8″	20′ 0″.

(Actual length of wall is measured from outside edge to outside edge of units and is equal to the nominal length minus ⅜″ (one mortar joint).)

Table 8–2. Nominal Height of Concrete Masonry Walls by Courses

No. of courses	Nominal height of concrete masonry walls	
	Units 7⅝″ high and ⅜″ thick bed joint	Units 3⅝″ high and ⅜″ thick bed joint
1	8″	4″.
2	1′ 4″	8″.
3	2′ 0″	1′ 0″.
4	2′ 8″	1′ 4″.
5	3′ 4″	1′ 8″.
6	4′ 0″	2′ 0″.
7	4′ 8″	2′ 4″.
8	5′ 4″	2′ 8″.
9	6′ 0″	3′ 0″.
10	6′ 8″	3′ 4″.
15	10′ 0″	5′ 0″.
20	13′ 4″	6′ 8″.
25	16′ 8″	8′ 4″.
30	20′ 0″	10′ 0″.
35	23′ 4″	11′ 8″.
40	26′ 8″	13′ 4″.
45	30′ 0″	15′ 0″.
50	33′ 4″	16′ 8″.

(For concrete masonry units 7⅝″ and 3⅝″ in height laid with ⅜″ mortar joints. Height is measured from center to center of mortar joints.)

Subsurface Drainage

When the ground water level in the wet season can be expected to be at the elevation of the basement floor, a line of drain tile should be placed in the outer side of footings. The tile line should have a fall of at least ½ inch in 12 feet and should drain to a suitable outlet. Pieces of roofing felt placed over the joints prevent sediment from entering the tile during backfilling. The tile line should be covered to a depth of 12 inches with a permeable fill of coarse gravel or crushed stone ranging from 1 to 1½ inches in size. Then the rest of the trench can be filled with earth from the excavation after the first floor is in place.

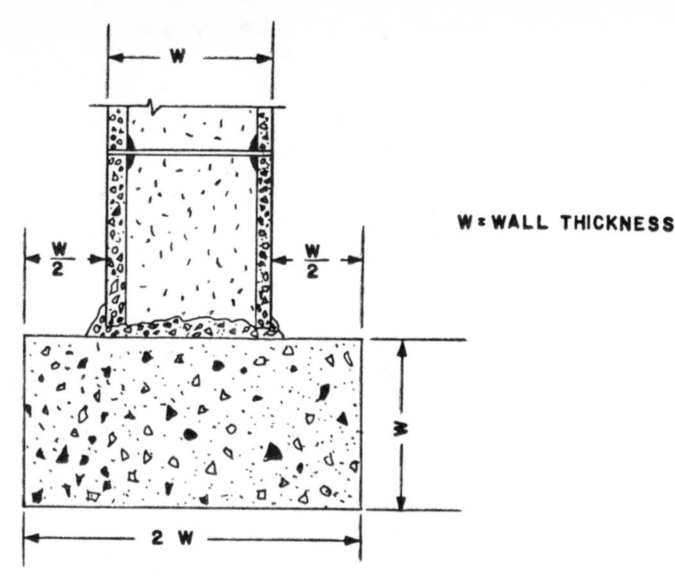

Figure 8–3. Dimensions of masonry wall footings.

Weathertight Walls

Construction Details. Good workmanship is always an important factor in building weathertight walls. Each masonry unit should be laid plumb and true. Both horizontal and vertical joints should be well filled and compacted by tooling when the mortar has partly stiffened. Flashing is necessary at vertical joints in copings and caps, at the joints between roofs and walls, and below cornices and other members projecting beyond the face of the wall. Drips should be provided for chimney caps, sills, and other projecting ledges to shed water away from the wall surface. Drains and gutters must be large enough to keep water from overflowing and running down over masonry surfaces.

Exterior Masonry Joints. Concave and V-shaped mortar joints (fig. 8–4) are recommended for walls of exterior concrete masonry in preference to struck or raked joints that form small lodges which may hold water. With modular-size

Table 8-3. Weights and Quantities of Materials for Concrete Masonry Walls

Actual unit sizes (width x height x length) in.	Nominal wall thickness in.	For 100 sq ft of wall			For 100 concrete units	
		Number of units	Average weight of finished wall		Mortar** cu ft	Mortar*** cu ft
			Heavyweight aggregate lb*	Lightweight aggregate lb*		
3⅝ x 3⅝ x 15⅝	4	225	3050	2150	13.5	6.0
5⅝ x 3⅝ x 15⅝	6	225	4550	3050	13.5	6.0
7⅝ x 3⅝ 15⅝	8	225	5700	3700	13.5	6.0
3⅝ x 7⅝ x 15⅝	4	112.5	2850	2050	8.5	7.5
5⅝ x 7⅝ x 15⅝	6	112.5	4350	2950	8.5	7.5
7⅝ x 7⅝ x 15⅝	8	112.5	5500	3600	8.5	7.5
11⅝ x 7⅝ x 15⅝	12	112.5	7950	4900	8.5	7.5

Table based on ⅜-in. mortar joints.
*Actual weight within ±7% of average weight.
**Actual weight within ±17% of average weight.
***With face-shell mortar bedding. Mortar quantities include 10% allowance for waste.
Actual weight of 100 sq ft of wall can be computed by formula W (N) + 150 (M) where:
W = actual weight of a single unit
N = number of units for 100 sq ft of wall
M = cu ft of mortar for 100 sq ft of wall

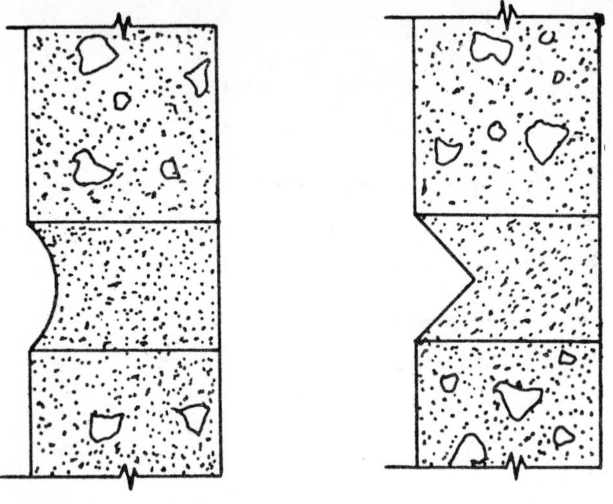

Figure 8–4. Tooled mortar joints for weathertight construction.

masonry units mortar joints will be approximately 3/8-inch thick. Experience has shown that this thickness of joint where properly made helps to produce a weathertight, neat, and durable concrete masonry wall.

Basement Walls

Plaster Coat. The earth side of concrete masonry basement walls should always be given two 1/4-inch thick coats of plaster. Either portland cement plaster (1—2 1/2 mix by volume) or mortar used in laying the block should be used for this purpose. In hot dry weather, the wall surface should be very lightly dampened with a fog spray of water before application of the first coat of plaster. The first coat of plaster is roughened after it has partly hardened to provide a bond for the second coat, and allowed to harden 24 hours before the second coat is applied. The first coat is dampened lightly just before the second coat is applied and the second coat should be kept damp for at least 48 hours after application. In very wet soils plastered surfaces below grade are frequently given two continuous coatings of bituminous material brushed on over a suitable priming coat. The plaster must be dry when the primer is applied to it and the primer coat must be dry when the bituminous material is applied. No backfilling against concrete masonry walls should be permitted until the first floor is in place.

Supporting Floor and Roof Loads. Masonry courses which support floor beams or floor slabs should be of solid masonry. This helps to distribute the loads over the wall and provides a barrier against termites. Such courses can be constructed by filling the cores of hollow block with concrete or mortar or by using solid masonry units without cores. Strips of expanded metal lath, laid in the bed joint below, support the concrete or mortar filling in the cores.

Concrete Masonry Walls

First Course. After locating the corners of the wall, the mason usually checks the layout by stringing out the blocks for the first course without mortar (1, fig. 8–5). A chalked snapline is useful to mark the footing and aline the block accurately. A full bed of mortar is then spread and furrowed with the trowel to insure plenty of mortar along the bottom edges of the face shells of the block for the first course (2, fig. 8–5). The corner block should be laid first and carefully positioned (3, fig. 8–5). All blocks should be laid with the thicker end of the face shell up to provide a larger mortar-bedding area (4, fig. 8–5). Mortar is applied only to the ends of the face shells for vertical joints. Several blocks can be placed on end and the mortar applied to the vertical face shells in

1 Stringing out blocks

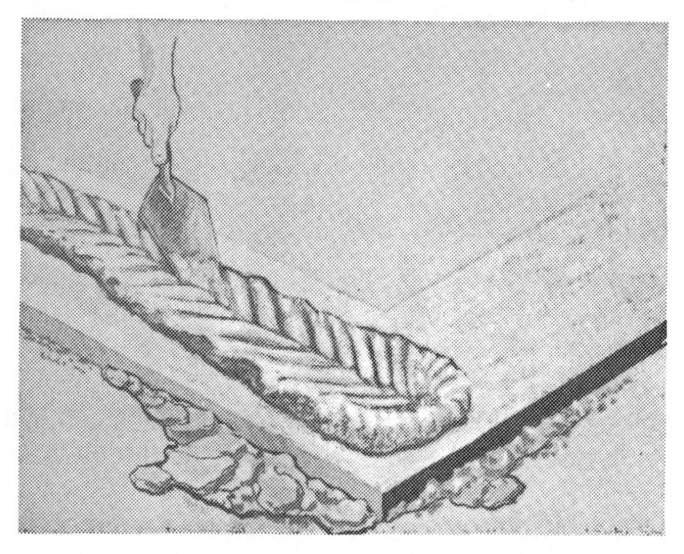

2 Spread and furrow mortar bed

important. It should be tipped slightly towards the mason so he can see the edge of the course below, enabling him to place the lower edge of the lock directly over the course below (fig. 8–11). All adjustments to final position must be made while the mortar is soft and plastic. Any adjustments made

3 Position corner block

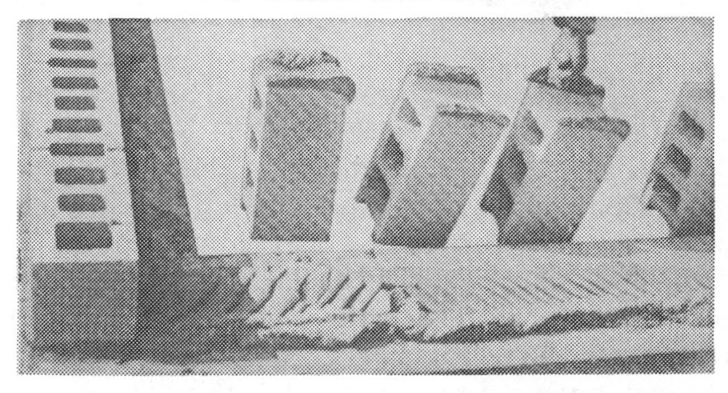

4 Blocks buttered for vertical joints

5 Positioning block

Figure 8–5. First course of blocks.

one operation. Each block is then brought over its final position and pushed downward into the mortar bed and against the previously laid block to obtain a well-filled vertical mortar joint (5, fig. 8–5). After three or four blocks have been laid, the mason's level is used as a straightedge to assure correct alinement of the blocks. Then the blocks are carefully checked with the level and brought to proper grade and made plumb by tapping with the trowel handle (1 and 2, fig. 8–6). The first course of concrete masonry should be laid with great care, to make sure it is properly alined, leveled, and plumbed, and to assure that succeeding courses, and finally the wall, are straight and true.

Laying Up the Corners. After the first course is laid, mortar is applied only to the horizontal face shells of the block (faceshell mortar bedding). Mortar for the vertical joints may be applied to the vertical face shells of the block to be placed or to the block previously laid or both, to insure well-filled joints (fig. 8–7). The corners of the wall are built first, usually four or five courses higher than the center of the wall. As each course is laid at the corner, it is checked with a level for alinement (1, fig. 8–8), for level (2, fig. 8–8), and plumb (3, fig. 8–8). Each block is carefully checked with a level or straightedge to make certain that the faces of the block are all in the same plane to insure true, straight walls. The use of a story or course-pole, a board with markings 8 inches apart, provides an accurate method of determining the top of the masonry for each course (fig. 8–9). Joints are $3/8$-inch thick. Each course, in building the corners, is stepped back a half block and the mason checks the horizontal spacing of the block by placing his level diagonally across the corners of the block (fig. 8–10).

1 Leveling block

2 Plumbing block

Figure 8–6. Checking first course of blocks.

Laying Block Between Corners. When filling in the wall between the corners, a mason's line is stretched from corner to corner for each course and the top outside edge of each block is laid to this line. The manner of gripping the bock is

Figure 8–7. Vertical joints.

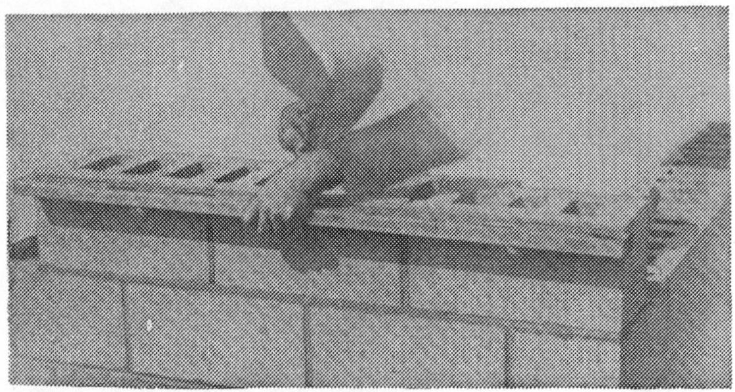

1 Aligning

3 Plumbing

2 Leveling

Figure 8–8. Checking each course.

Figure 8–9. Use of story or course pole.

Figure 8–10. Checking horizontal spacing of blocks.

after the mortar has stiffened will break the mortar bond and allow the penetration of water. Each block is leveled and alined to the mason's line by tapping lightly with the trowel handle. The use of the mason's level between corners is limited to checking the face of each block to keep it lined up with the face of the wall.

Mortar. To assure good bond, mortar should not be spread too far ahead of actual laying of the block or it will stiffen and lose its plasticity. As each block is laid, excess mortar extruding from the joints is cut off with the trowel (fig. 8–12) and is thrown back on the mortar board to be reworked into the fresh mortar. Dead mortar that has been picked up from the scaffold or from the floor should not be used.

Closure Block. When installing the closure block, all edges of the opening and all four vertical edges of the closure block are buttered with mortar and the closure block is carefully lowered into place (fig. 8–13). If any of the mortar falls out leaving an open joint, the block should be removed and the procedure repeated.

Tooling. Weathertight joints and neat appearance of concrete block walls are dependent on proper tooling. The mortar joints should be tooled after a section of the wall has been laid and the mortar has become "thumb-print" hard. Tooling (fig. 8–14) compacts the mortar and forces it tightly against the masonry on each side of the joint. All joints should be tooled either concave or V-shaped. Horizontal joints (1, fig. 8–14) should be tooled first, followed by striking the vertical joints with a small S-shaped jointer (2, fig. 8–14). Mortar burrs remaining after tooling is completed should be trimmed off flush with the face of the wall with a trowel or removed by rubbing with a burlap bag.

Figure 8–11. Adjusting block between corners.

Figure 8–12. Cutting off excess mortar.

Figure 8–13. Installing closure block.

Anchor Bolts. Wood plates are fastened to tops of concrete masonry walls by anchor bolts ½ inch in diameter, 18 inches long and spaced not more than 4 feet apart. The bolts are placed in cores of the top two courses of block with the cores filled with concrete or mortar. Pieces of metal lath placed in the second horizontal mortar joint from the top of the wall and under the cores to be filled (1, fig. 8–15) will hold the concrete or mortar filling in place. The threaded end of the bolt should extend above the top of the wall, (2, fig. 8–15).

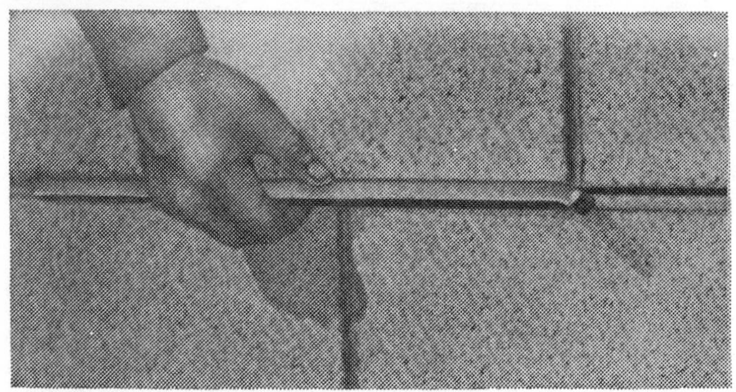

1 Tooling horizontal joints

2 Striking vertical joints

Figure 8–14. Tooling mortar joints.

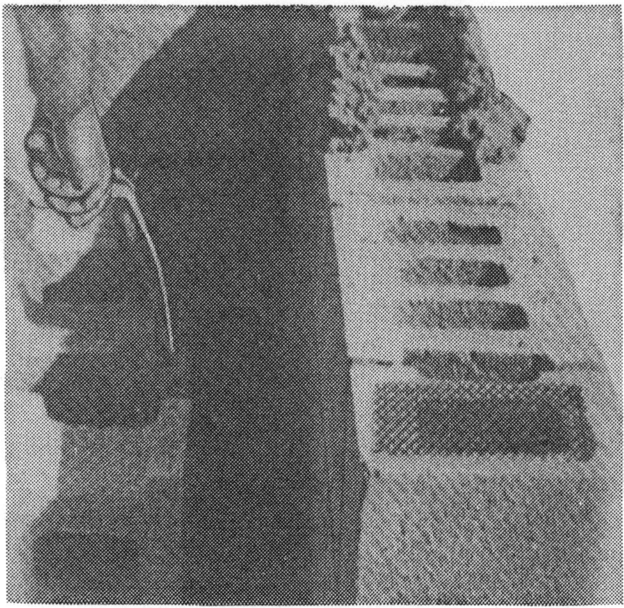

1 Placing metal lath

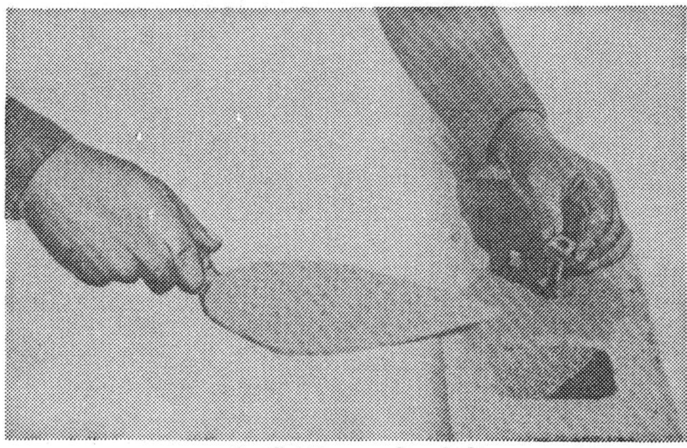

2 Setting anchor bolt

Figure 8–15. Installing anchor bolts on top of wall.

Control Joints. Control joints are continuous vertical joints built into concrete masonry walls to control cracking resulting from unusual stresses. The joints are intended to permit slight wall movement without cracking. Control joints should be laid up in mortar just as any other joint. Full-

and half-length block are used to form a continuous vertical joint (1, fig. 8–16). If they are exposed to the weather or to view, they should be calked. After the mortar is quite stiff, it should be raked out to a depth of about 3/4 inch to provide a recess for the calking material (2, fig. 8–16). A thin, flat calking trowel is used to force the calking compound into the joint. Another type of control joint can be constructed with building paper or roofing felt inserted in the end core of the block and extending the full height of the control joint (fig. 8–17). The paper or felt, cut to convenient lengths and wide enough to extend across the joint, prevents the mortar from bonding on one side of the joint. Sometimes control joint blocks are used if available.

Intersecting Walls

Bearing Walls. Intersecting concrete block bearing walls should not be tied together in a masonry bond, except at the corners. Instead, one wall should terminate at the face of the other wall with a control joint at the point. Bearing walls are tied together with a metal tiebar 1/4 × 1 1/4 × 28 inches, with 2-inch right angle bends on each end (1, fig. 8–18). Tiebars are spaced not over 4 feet apart vertically. Bends at the ends of the tiebars are embedded in cores filled with mortar or concrete (2, fig. 8–18). Pieces of metal lath placed under the cores support the concrete or mortar filling (1, fig. 8–15).

Nonbearing Walls. To tie nonbearing block walls to other walls, strips of metal lath or 1/4-inch mesh galvanized hardware cloth are placed across the joint between the two walls (1, fig. 8–19), in alternate courses in the wall. When one wall is constructed first, the metal strips are built into the wall and later tied into the mortar joint of the

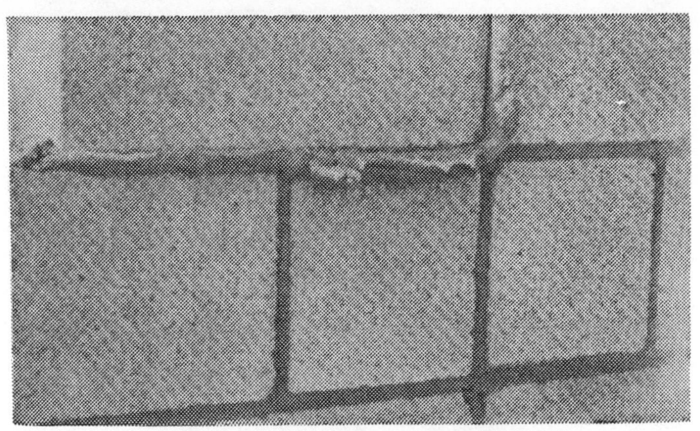

1 Full and half length block for joint

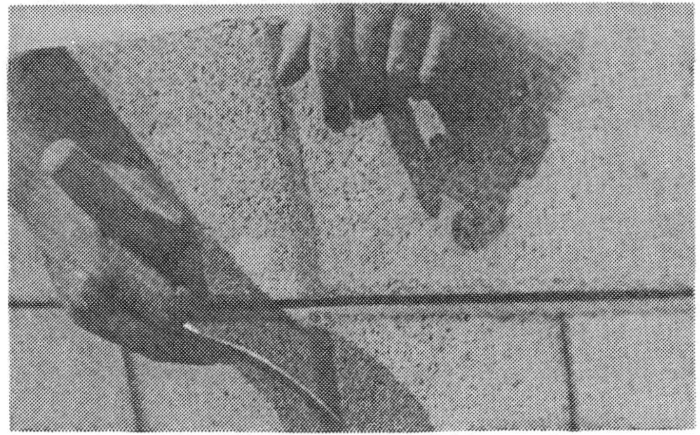

2 Raking mortar from joint

Figure 8–16. Control joint.

second wall (2, fig. 8–19). Control joints are constructed where the two walls meet.

Lintels and Sills

Lintels. Precast concrete lintels are often used over door and window openings (1, fig. 8–20). Precast concrete lintels are designed with an offset on the underside (2, fig. 8–20) for modular window and door openings. Steel lintel angles are also used for lintels to support block over openings. They must be installed with an offset on

Figure 8–17. Paper or felt used for control joints.

the underside (3, fig. 8–20) to fit modular openings. A non-corroding metal plate placed under the ends of lintels where control joints occur, will permit lintels to slip and the control joints to function properly. A full bed of mortar should be placed over this metal plate to distribute the lintel load uniformly.

Sills. Precast concrete sills are usually installed after the masonry walls have been built (fig. 8–21). Joints at the ends of the sills should be tightly filled with mortar or with a calking compound.

Patching and Cleaning Block Walls

Patching. Any patching of the mortar joints or filling of holes left by nails or line pins should be done with fresh mortar.

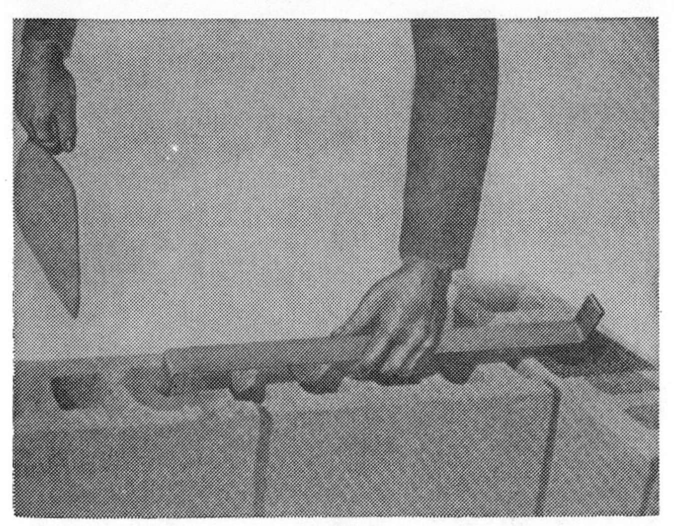

1 Tiebar

2 Filling core with mortar

Figure 8–18. Tieing intersecting bearing walls.

Cleaning. Hardened, embedded mortar smears, cannot be removed and paint cannot be depended on to hide smears, so particular care should be taken to prevent smearing mortar into the surface of the block. Concrete block walls should not be cleaned with an acid wash to remove smears or mortar droppings. Mortar droppings that stick to the block wall should be allowed to dry before removal with a trowel (1, fig. 8–22).

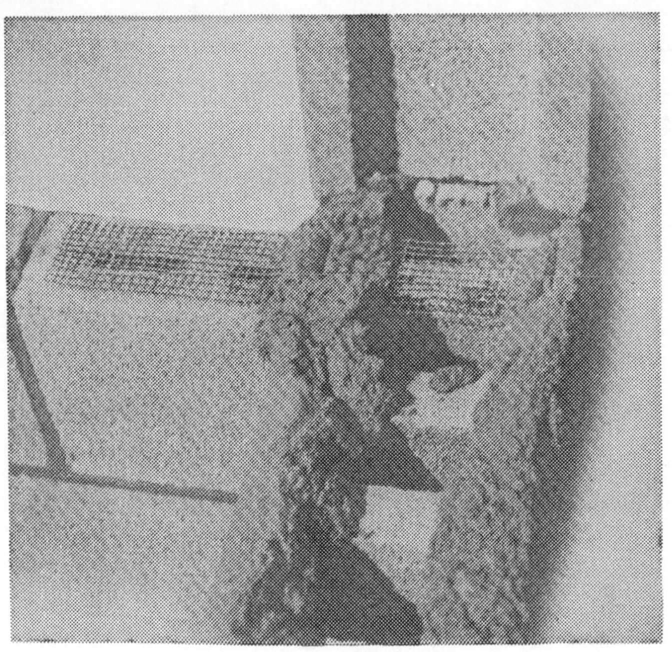

1 Use of metal lath

2 Mortar joint between walls

Figure 8–19. Tieing intersecting nonbearing walls.

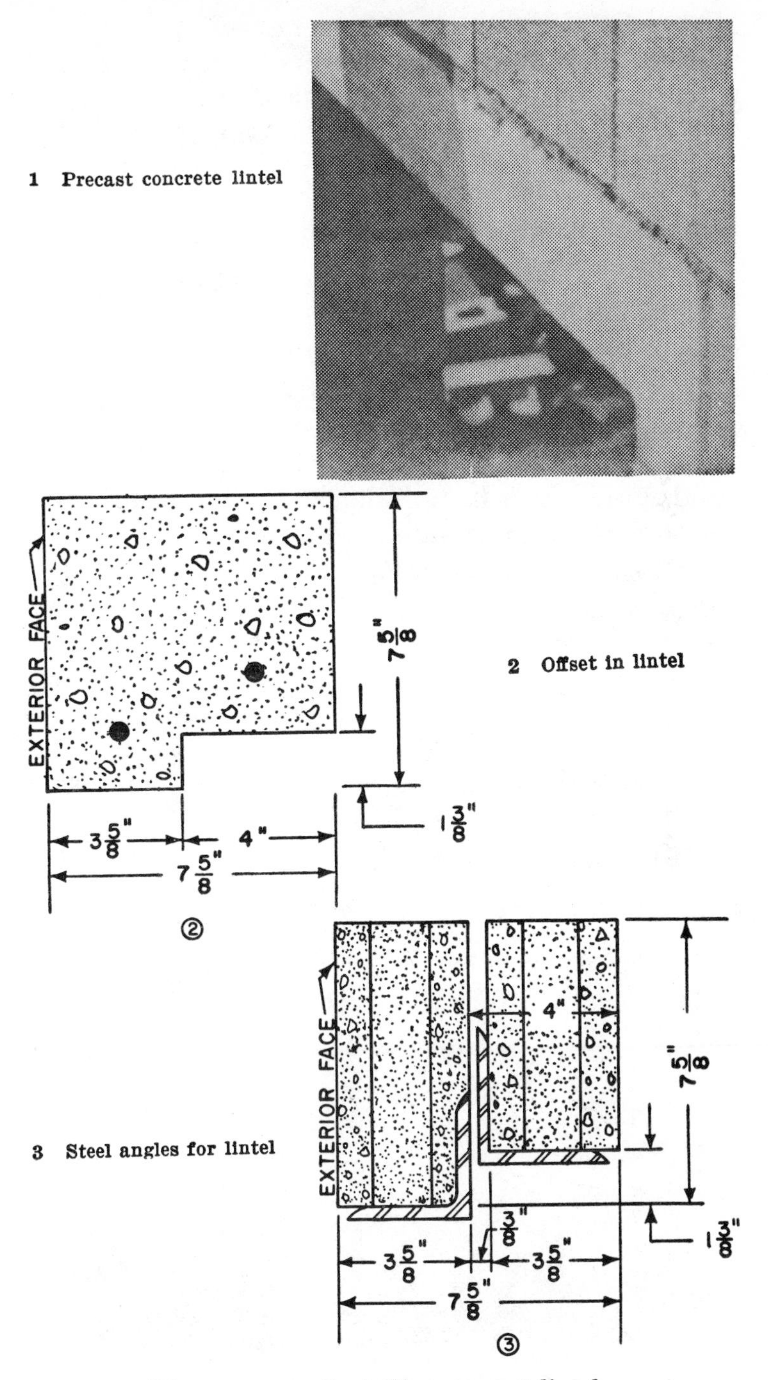

Figure 8–20. Installing precast lintels.

Most of the mortar can be removed by rubbing with a small piece of concrete (broken) block after the mortar is dry and hard (2, fig. 8–22). Brushing the rubbed spots will remove practically all of the mortar (3, fig. 8–22).

Duties of Concrete Mason and Helper

Mason. The mason is responsible for laying out the job so that the finished masonry will be properly done. If the construction involves walls, he must see that the walls are plumb and the courses are level. He is responsible for all the detail work such as cutting and fitting of masonry units, joints, and installation of anchor bolts and ties for intersecting walls.

Mason's Helper. The mason's helper mixes mortar and carries concrete blocks and mortar to the mason as rapidly as these materials are required. He assists the mason in the layout of the job and at times he may lay out block on an adjacent course to expedite the mason's work. He keeps the mortar tempered as required.

RUBBLE STONE MASONRY

Uses

Rubble stone masonry such as that shown in figure 8–23 is used for walls both above and below

Figure 8–21. Installing precast concrete sills.

1 Removing mortar with trowel

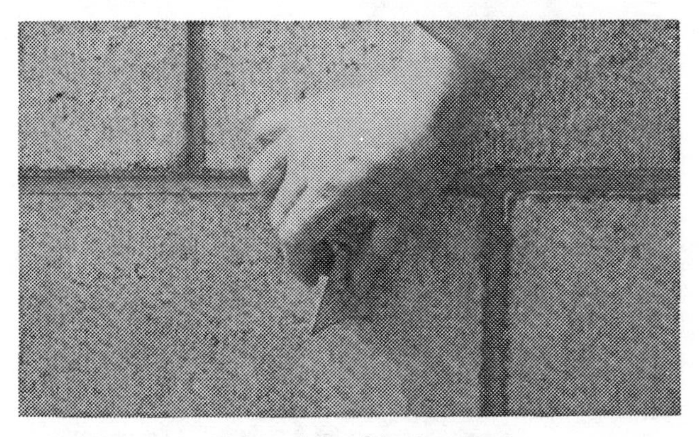

2 Using piece of broken block

3 Brushing

Figure 8–22. Patching and cleaning concrete block.

1 Random rubble masonry

2 Coursed rubble masonry

Figure 8–23. Rubble stone masonry.

ground and for bridge abutments. In military construction it is used when form lumber or masonry units are not available. Rubble masonry may be laid up with or without mortar; if strength and stability are desired, mortar must be used.

Types

Random Rubble. This is the crudest of all types of stonework. Little attention is paid to laying the stone in courses (1, fig. 8–23). Each layer must contain bonding stones that extend through the wall (fig. 8–24). This produces a wall that is well tied together. The bed joints should be horizontal for stability but the "builds" or head joints may run in any direction.

Coursed Rubble. Coursed rubble is assembled of roughly squared stones in such a manner as to produce approximately continuous horizontal bed joints (2, fig. 8–23).

Materials for Use in Random Rubble Stone Masonry

Stone. The stone for use in random rubble stone masonry should be strong, durable, and cheap. Durability and strength depend upon the chemical composition and physical structure of the

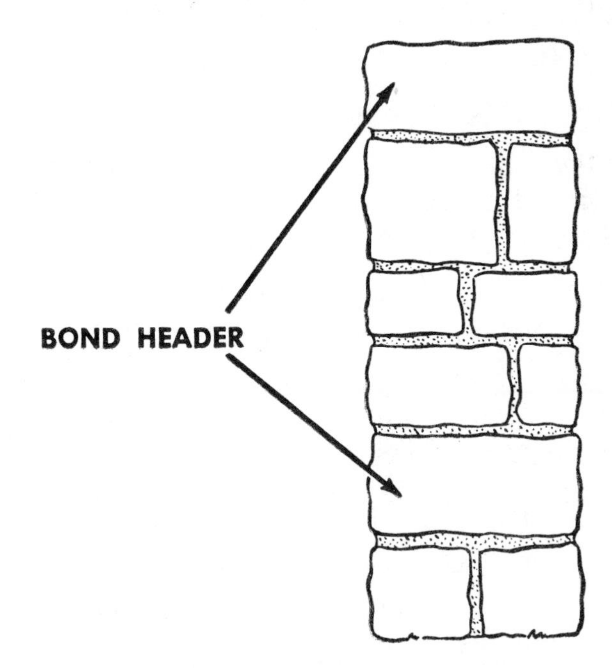

Figure 8–24. Rubble stone masonry wall showing bonding stone.

stone. Some of the more commonly found stones that are suitable are limestone, sandstone, granite, and slate. Unsquared stones obtained from nearby ledges or quarries or even field stones may be used. The size of the stone should be such that two men can easily handle it. A variety of sizes is necessary in order to avoid using large quantities of mortar.

Mortar. The mortar for use in random rubble masonry may be composed of portland cement and sand in the proportions of one part cement to three parts sand by volume. Such mortar shrinks excessively and does not work well with the trowel. A better mortar to use is portland-cement-lime mortar (table 7–1). Mortar made with ordinary portland cement will stain most types of stone. If staining must be prevented, nonstaining white portland cement should be used in making the mortar. Lime does not usually stain the stone.

Laying Rubble Stone Masonry

Workmanship in laying stone masonry affects the economy, durability, and strength of the wall more than any other factor.

Rules for Laying.

Each stone should be laid on its broadest face.

If appearance is to be considered, the larger stones should be placed in the lower courses. The size of the stones should gradually diminish toward the top of the wall.

Porous stones should be moistened before being placed in mortar in order to prevent the stone from absorbing water from the mortar and thereby weakening the bond between the stone and the mortar.

The spaces between adjoining stones should be as small as practicable and these spaces should be completely filled with mortar and smaller stones.

If necessary to remove a stone after it has been placed upon the mortar bed, it should be lifted clear and reset.

Footing. The footing is larger than the wall itself. The largest stones should be used in it to give the greatest strength and lessen the danger of unequal settlement. The footing stones should be as long as the footing is wide, if possible. The footing stones should be laid in a mortar bed about 2 inches deep and all space between the stones filled with mortar and smaller stones.

Bed Joint. The thickness of the bed joint will vary, depending upon the stone used. In making the bed joint, enough mortar should be spread on the stone below the one being placed to fill the space between the two stones completely. Care must be taken not to spread the mortar too far ahead of the stonelaying.

Head Joint or Builds. The head joint is made after three or four stones have been laid. This is done by slushing the small spaces with mortar and filling the larger spaces with small stones and mortar. The head joints should be formed before the mortar in the bed joint has set up.

Bonding. Bond stones should occur at least once in each 6 to 10 square feet of wall. These stones pass all the way through the wall as shown in figure 8–24. Each head joint should be offset from adjacent head joints above and below it as much as possible (1, fig. 8–23) to bond the wall together and make it stronger.

Laying the Wall.

(1) If the wall need not be exactly plumb and true to line. the level and line will not be used and

the wall will be laid by eye. Frequent sighting is necessary.

If the wall must be exactly plumb and erected to line, corner posts of wood should be erected to serve the purpose of corner leads and the stone laid with a line. No particular attention is paid to laying the stone in level courses. Some parts of the stone will be farther away from the line than other parts.

9

Brick And Tile Masonry

CHARACTERISTICS OF BRICK AND BRICK MASONRY

Definition. Brick masonry is that type of construction in which units of baked clay or shale of uniform size, small enough to be placed with one hand, are laid in courses with mortar joints to form walls of virtually unlimited length and height. Brick are kiln-baked from various clay and shale mixtures. The chemical and physical characteristics of the ingredients vary considerably; these and the kiln temperatures combine to produce brick in a variety of colors and hardnesses. In some regions, pits are opened and found to yield clay or shale which, when ground and moistened, can be formed and baked into durable brick; in other regions, clays or shales from several pits must be mixed.

Brick Sizes. Standard bricks manufactured in the United States are 2¼ by 3¾ by 8 inches. English bricks are 3 by 4½ by 9 inches, Roman bricks are 1½ by 4 by 12 inches, and Norman bricks are 2¾ by 4 by 12 inches. The actual dimensions of brick vary a little because of shrinkage during burning.

Cut Brick. Frequently the bricklayer cuts the brick into various shapes. The more common of these are shown in figure 9–1. They are called half

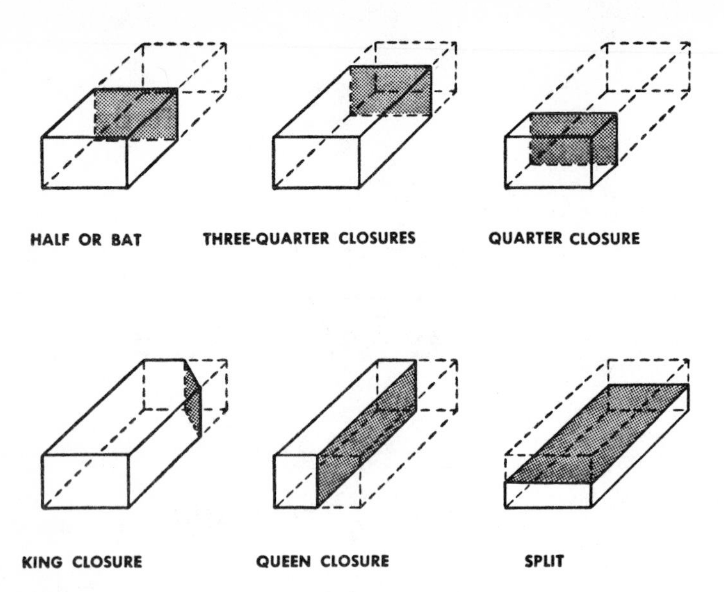

Figure 9–1. Shapes of cut brick.

or bat, three-quarter closure, quarter closure, king closure, queen closure, and split. They are used to fill in the spaces at corners and such other places where a full brick will not fit.

Names of Brick Surfaces. The six surfaces of a brick are called the face, the side, the cull, the end, and the beds, as shown in figure 9–2.

Brick Classification. There are three general types of structural clay masonry units: solid ma-

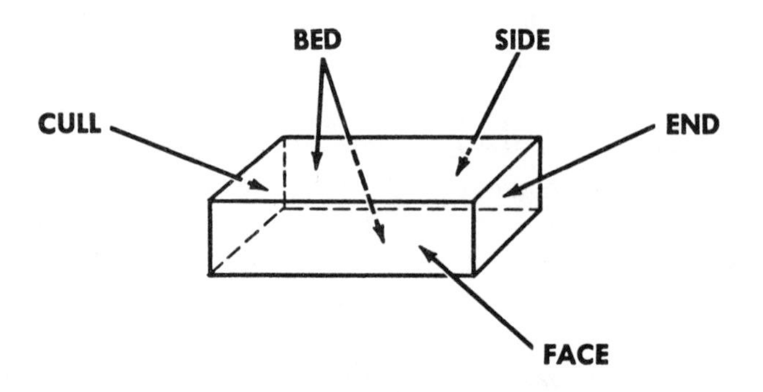

Figure 9–2. Names of brick surfaces.

sonry units, hollow masonry units and architectural terra cotta. These units may serve a structural function only, as a decorative finish, or a combination of both. Structural clay products include brick, hollow tile of all types and architectural terra cotta. They do not include thin wall tile, sewer pipe, flue linings, drain tile and the like.

Types of Bricks. There are many types of brick. Some are different in formation and composition while others vary according to their use. Some commonly used types of bricks are:

Building brick. The term building brick, formerly called common brick, is applied to brick made of ordinary clays or shales and burned in the usual manner in the kilns. These bricks do not have special scorings or markings and are not produced in any special color or surface texture. Building brick is also known as hard and kiln run brick. It is used generally for the backing courses in solid or cavity brick walls. The harder and more durable kinds are preferred for this purpose.

Face brick. Face brick are used in the exposed face of a wall and are higher quality units than backup brick. They have better durability and appearance. The most common colors of face brick are various shades of brown, red, gray, yellow, and white.

Clinker brick. When bricks are overburned in the kilns, they are called clinker brick. This type of brick is usually hard and durable and may be irregular in shape. Rough hard corresponds to the clinker classification.

Pressed brick. The dry press process is used to make this class of brick which has regular smooth faces, sharp edges, and perfectly square corners. Ordinarily all press brick are used as face brick.

Glazed brick. This type of brick has one surface of each brick glazed in white or other color. The ceramic glazing consists of mineral ingredients which fuse together in a glass-like coating during burning. This type of brick is particularly suited for walls or partitions in hospitals, dairies, laboratories or other buildings where cleanliness and ease of cleaning is necessary.

Fire brick. This type of brick is made of a special type of fire clay which will withstand the high temperatures of fireplaces, boilers and similar usages without cracking or decomposing. Fire brick is generally larger than regular structural brick and often it is hand molded.

Cored brick. Cored brick are brick made with two rows of five holes extending through their beds to reduce weight. There is no significant difference between the strength of walls constructed with cored brick and those constructed with solid brick. Resistance to moisture penetration is about the same for both types of walls. The most easily available brick that will meet requirements should be used whether the brick is cored or solid.

European brick. The strength and durability of most European clay brick, particularly English and Dutch, compares favorably with the clay brick made in the United States.

Sand-lime brick. Sand-lime bricks are made from a lean mixture of slaked lime and fine silicious sand molded under mechanical pressure and hardened under steam pressure. They are used extensively in Germany.

Strength of Brick Masonry

The principal factors governing the strength of brick masonry are:

Strength of the brick.

Strength and elasticity of the mortar.

Workmanship of the bricklayer.

Uniformity of the brick used.

Method used in laying the brick.

The strength of an individual brick varies widely, depending upon the material and manufacturing method. Brick with ultimate compressive strengths as low as 1,600 pounds per square inch have been made, whereas some well-burned bricks have compressive strengths exceeding 15,000 pounds per square inch.

The strength of portland-cement-lime mortar is normally higher than the strength of the brick. Because of this, the strength of brick masonry laid in cement-lime mortar is higher than the strength of the individual brick. The use of plain lime mortar reduces the load-carrying capacity of a wall or column to considerably less than half the load-carrying capacity of the same type construction in which portland-cement-lime mortar has been used. The compressive working strength of a brick wall or column that has been laid up with cement-lime mortar is normally from 500 to 600 pounds per square inch.

High suction brick, if laid dry, will absorb water from the mortar before the bond has developed and, for this reason, such brick must be thoroughly wetted before laying to obtain good masonry construction. In wetting brick, sprinkling is not sufficient. A hose stream should be played on the pile until water runs from all sides, after which the brick should be allowed to surface dry before they are laid. Water on the surface of the brick will cause floating on the mortar bed. A rough but effective on-the-job test for determining whether the brick should be wet before laying is to sprinkle a few drops of water on the flat side of the brick.

If these drops of water are absorbed completely in less than 1 minute, the brick should be wet when laid.

Resistance to Weathering

The resistance of masonry walls to weathering depends almost entirely upon their resistance to water penetration because freezing and thawing action is virtually the only type of weathering that affects brick masonry. With the best workmanship, it is possible to build brick walls that will resist the penetration of rain water during a storm lasting as long as 24 hours accompanied by a 50- to 60-mile-per-hour wind. In most construction, it is unreasonable to expect the type of workmanship required to build a wall that will allow no water pentration. It is advisable to provide some means of taking care of moisture after it has penetrated the brick masonry. Properly designed flashing and cavity walls are two ways of handling moisture that has entered the wall.

Important factors in preventing the entrance of water are tooled mortar joints and caulking around windows and door frames.

The joints between the brick must be solidly filled, especially in the face tier. Slushing or grouting the joints after the brick has been laid does not completely fill the joint. The mortar joint should be tooled to a concave surface before the mortar has had a chance to set up. In tooling, sufficient force should be used to press the mortar tight against the brick on both sides of the mortar joint.

Mortar joints that are tightly bonded to the brick have been shown to have greater resistance to moisture penetration than joints not tightly bonded to the brick.

Fire Resistance

Fire-resistance tests conducted upon brick walls laid up with portland-cement-lime mortar have made it possible to give fire-resistance periods for various thicknesses of brick walls. A summary is given in table 9–1. The tests were made using the American Society for Testing Materials standard method for conducting fire tests.

General characteristics of Brick Masonry

Heat-Insulating Properties. Solid brick masonry walls provide very little insulation against heat and cold. A cavity wall or a brick wall backed with hollow clay tile has much better insulating value.

Sound-Insulating Properties. Because brick walls are exceptionally massive, they have good sound-insulating properties. In general, the heavier the wall, the better will be its sound-insulating value; however, there is no appreciable increase in sound insulation by a wall more than 12 inches thick as compared to a wall between 10 and 12 inches thick. The expense involved in constructing a thicker wall merely to take advantage of the slight increase is too great to be worthwhile. Dividing the wall into two or more layers, as in the case of a cavity wall, will increase its resistance to the transmission of sound from one side of the wall to the other. Brick walls are poor absorbers of sound originating within the walls and reflect much of it back into the structure. Sounds caused by impact, as when the wall is struck with a hammer, will travel a great distance along the wall.

Expansion and Construction. Brick masonry expands and contracts with temperature change. Walls up to a length of 200 feet do not need expansion joints. Longer walls need an expansion joint for every 200 feet of wall. The joint can be made

Table 9–1. ***Fire Resistance of Brick Load-Bearing Walls Laid with Portland Cement Mortar***

Normal wall thickness (inches)	Type of wall	Material	Ultimate fire-resistance period. Incombustible members framed into wall or not framed in members		
			No plaster (hours)	Plaster on one side* (hours)	Plaster on two sides* (hours)
4	Solid	Clay or shale	1¼	1¾	2½
8	Solid	Clay or shale	5	6	7
12	Solid	Clay or shale	10	10	12
8	Hollow rowlock	Clay or shale	2½	3	4
12	Hollow rowlock	Clay or shale	5	6	7
9 to 10	Cavity	Clay or shale	5	6	7
4	Solid	Sand-lime	1¾	2½	3
8	Solid	Sand-lime	7	8	9
12	Solid	Sand-lime	10	10	12

*Not less than ½ inch of 1:3 sanded gypsum plaster is required to develop these ratings.

as shown in figure 4–26. A considerable amount of the expansion and contraction is taken up in the wall itself. For this reason, the amount of movement that theoretically takes place does not actually occur.

Abrasion Resistance. The resistance of brick to abrasion depends largely upon it's compressive strength, related to the degree of burning. Well-burned brick have excellent wearing quantities.

Weight of Brick. The weight of brick varies from 100 to 150 pounds per cubic foot depending upon the nature of the materials used in making the brick and the degree of burning. Well-burned brick are heavier than under-burned brick.

Fundamentals

Good bricklaying procedure depends on good workmanship and efficiency. Means of obtaining good workmanship are treated below. Efficiency involves doing the work with the fewest possible motions. The bricklayer studies his own operations to determine those motions that are unnecessary. Each motion should have a purpose and should accomplish a definite result. After learning the fundamentals, every bricklayer develops his own methods for achieving maximum efficiency. The work must be arranged in such a way that the bricklayer is continually supplied with brick and mortar. The scaffolding required must be planned before the work begins. It must be built in such a way as to cause the least interference with other workmen. Masons' tools and equipment which are generally the same or similar to those used in bricklaying are discussed in paragraphs 7–1 and 7–2.

9–7. Types of Bonds

The word bond, when used in reference to masonry, may have three different meanings:

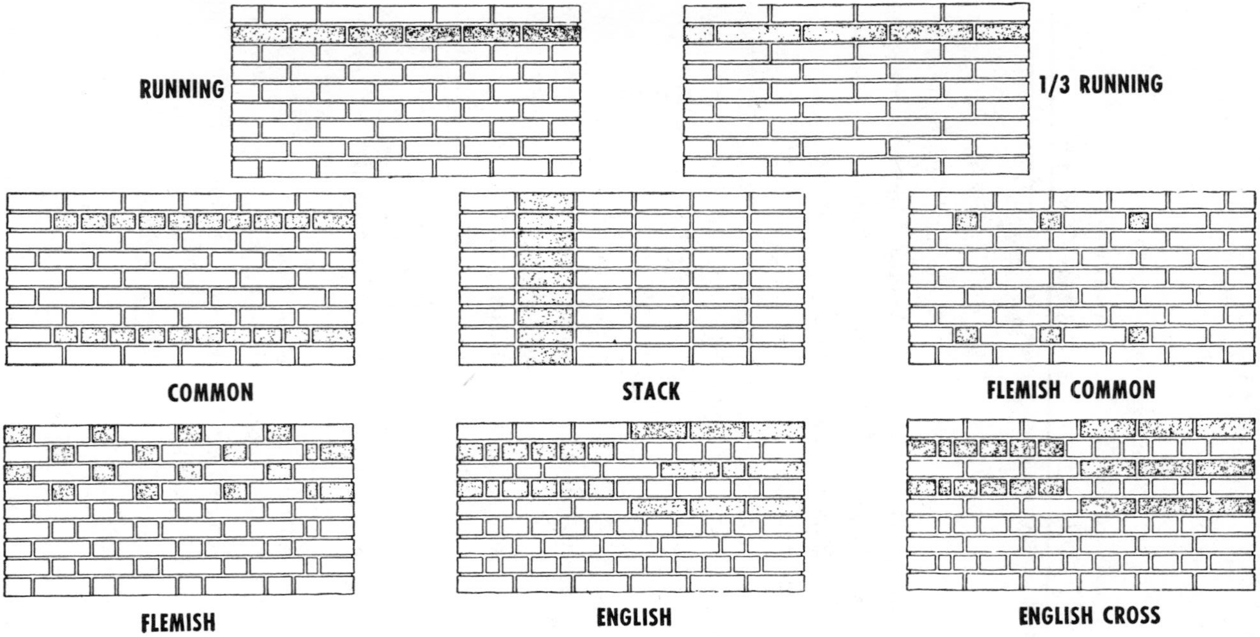

Figure 9–3. Some types of brick masonry bond.

Structural Bond. Structural bond is the method by which individual masonry units are interlocked or tied together to cause the entire assembly to act as a single structural unit. Structural bonding of brick and tile walls may be accomplished in three ways. First, by overlapping (interlocking) the masonry units, second by the use of metal ties imbedded in connecting joints, and third by the adhesion of grout to adjacent wythes of masonry.

Mortar Bond. Mortar bond is the adhesion of the joint mortar to the masonry units or to the reinforcing steel.

Pattern Bond. Pattern bond is the pattern formed by the masonry units and the mortar joints on the face of a wall. The pattern may result from the type of structural bond used or may be purely a decorative one in no way related to the structural bond. There are five basic pattern bonds in common use today (fig. 9–3): running bond, common or American bond, Flemish bond, English bond, and block or stack bond.

Running bond. This is the simplest of the basic pattern bonds, the running bond consists of all stretchers. Since there are no headers used in this bond, metal ties are usually used. Running bond is used largely in cavity wall construction and veneered walls of brick, and often in facing tile walls where the bonding may be accomplished by extra width stretcher tile.

Common or American bond. Common bond is a variation of running bond with a course of full length headers at regular intervals. These headers provide structural bonding as well as pattern. Header courses usually appear at every fifth, sixth, or seventh course depending on the structural bonding requirements. In laying out any bond pattern it is very important that the corners be started correctly. For common bond, a "three-quarter" brick must start each header course at the corner. Common bond may be varied by using a Flemish header course.

Flemish bond. Each course of brick is made up of alternate stretchers and headers, with the headers in alternate courses centered over the stretchers in the intervening courses. Where the headers are not used for the structural bonding, they may be obtained by using half brick, called "blind-headers." There are two methods used in starting the corners. Figure 9–3 shows the so called "Dutch" corner in which a three-quarter brick is used to start each course and the "English" corner in which 2 inch or quarter-brick closures must be used.

English bond. English bond is composed of alternate courses of headers and stretchers. The headers are centered on the stretchers and joints between stretchers in all courses line up vertically. Blind headers are used in courses which are not structural bonding courses.

Block or stack bond. Stack bond is purely a pattern bond. There is no overlapping of the units, all vertical joints being alined. Usually this pattern is bonded to the backing with rigid steel ties, but when 8 inch thick stretcher units are available, they may be used. In large wall areas and in load-bearing construction, it is advisable to reinforce the wall with steel pencil rods placed in the horizontal mortar joints. The vertical alinement requires dimensionally accurate units, or carefully prematched units, for each vertical joint alinement. Variety in pattern may be achieved by numerous combinations and modifications of the basic patterns shown.

English cross or Dutch bond. This bond is a variation of English bond and differs only in that vertical joints between the stretchers in alter-

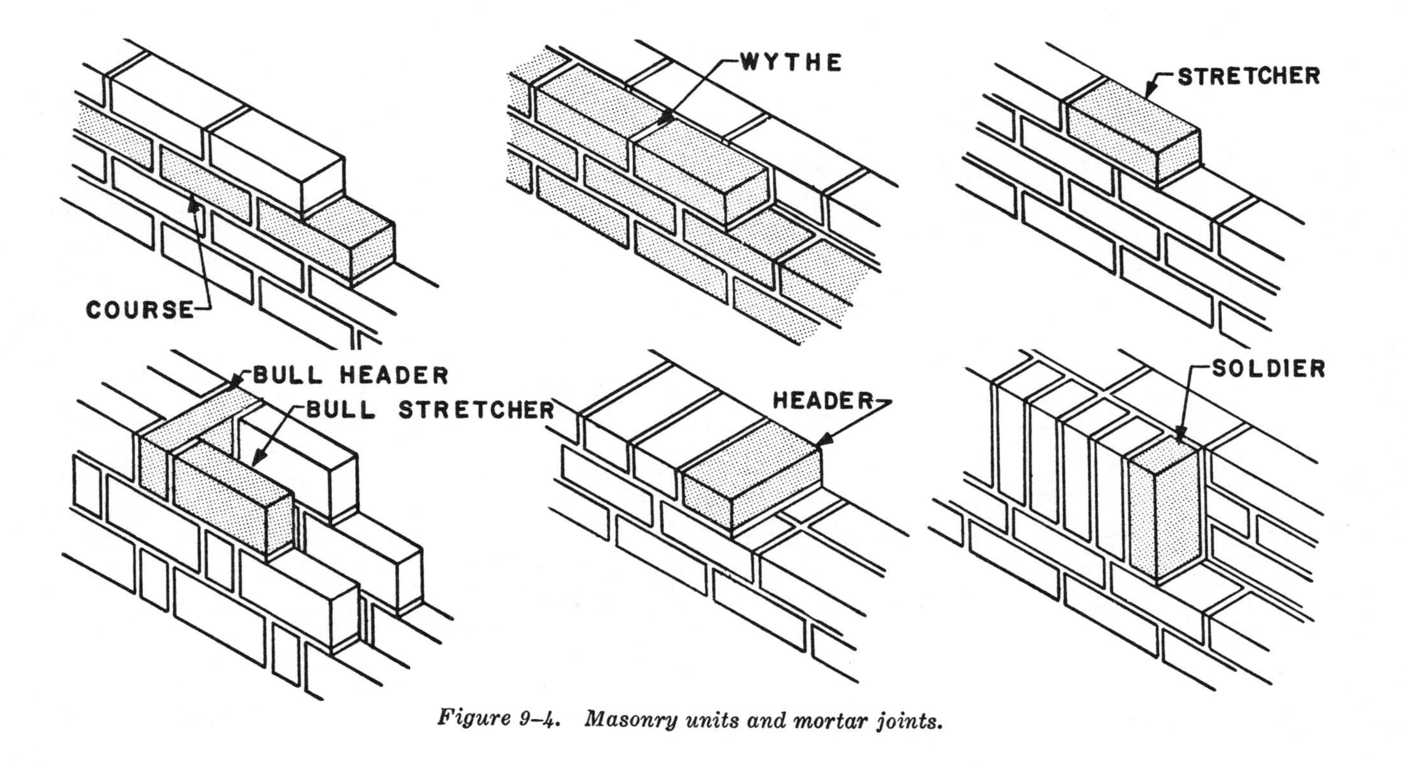

Figure 9–4. Masonry units and mortar joints.

nate courses do not line up vertically. These joints center on the stretchers themselves in the courses above and below.

Masonry Terms. Specific terms are used to describe the various positions of masonry units and mortar joints in a wall (fig. 9–4):

Course. One of the continuous horizontal layers (or rows) of masonry which, bonded together, form the masonry structure.

Wythe. A continuous vertical 4-inch or greater section or thickness of masonry as the thickness of masonry separating flues in a chimney.

Stretcher. A masonry unit laid flat with its longest dimension parallel to the face of the wall.

Header. A masonry unit laid flat with its longest dimension perpendicular to the face of the wall. It is generally used to tie two wythes of masonry together.

Rowlock. A brick laid on its edge (face).

Bull-Stretcher. A rowlock brick laid with its longest dimension parallel to the face of the wall.

Bull-Header. A rowlock brick laid with its longest dimension perpendicular to the face of the wall.

Soldier. A brick laid on its end so that its longest dimension is parallel to the vertical axis of the face of the wall.

Metal Ties. Metal ties can be used to tie the brick on the outside face of the wall to the backing courses. These are used when no header courses are installed. They are not as satisfactory as header courses. Typical metal ties are shown in figure 9–5.

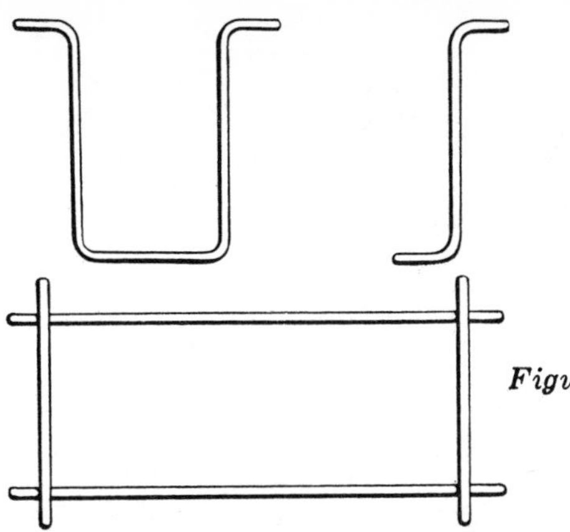

Figure 9–5. Metal ties.

Flashing. Flashing is installed in masonry construction to divert moisture, which may enter the masonry at vulnerable spots, to the outside. Flashing should be provided under horizontal masonry surfaces such as sills and copings, at intersections of masonry walls with horizontal surfaces such as roof and parapet or roof and chimney, over heads of openings such as doors and windows, and frequently at floor lines, depending upon the type of construction. To be most effective, the flashing should extend through the outer face of the wall and be turned down to form a drop. Weep holes should be provided at intervals of 18 inches to 2 feet to permit the water which accumulates on the flashing to drain to the outside. If, because of appearance, it is necessary to stop the flashing back of the face of the wall, weep holes are even more important than when the flashing extends through the wall. Concealed flashing with tooled mortar joints frequently will retain water in the wall for long periods and, by concentrating moisture at one spot, may do more harm than good.

Mortar Joints and Pointing

Holding the Trowel. The trowel should be held firmly in the position shown in figure 9–6.

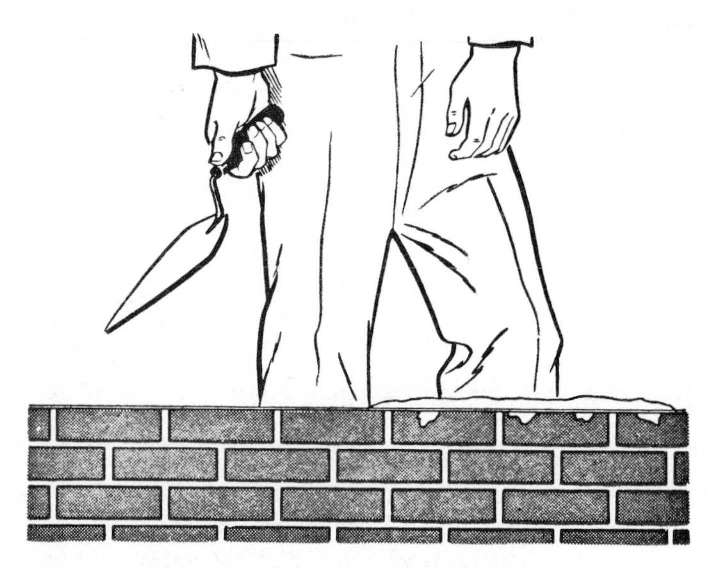

Figure 9–6. Correct way to hold a trowel.

The thumb should rest on top of the handle and should not encircle it.

Picking up Mortar. A right-handed bricklayer picks up mortar with the left edge of the trowel from the outside of the pile (1, fig. 9–7). He picks up the correct amount of spread for one to five bricks, according to the wall space and his skill. A pickup for one brick forms a small windrow along the left edge of the trowel. A pickup for five bricks is a full load for a large trowel (2, fig. 9–7).

Spreading Mortar. Holding the trowel with its left edge directly over the centerline of the previous course, the bricklayer tilts the trowel slightly and moves it to the right, dropping a windrow of mortar along the wall unit until the trowel is empty (3, fig. 9–7). In some instances mortar will be left on the trowel when the spreading of mortar on the course below has been completed. When this occurs the remaining mortar is returned to the board. A right- handed bricklayer works from left to right along the wall.

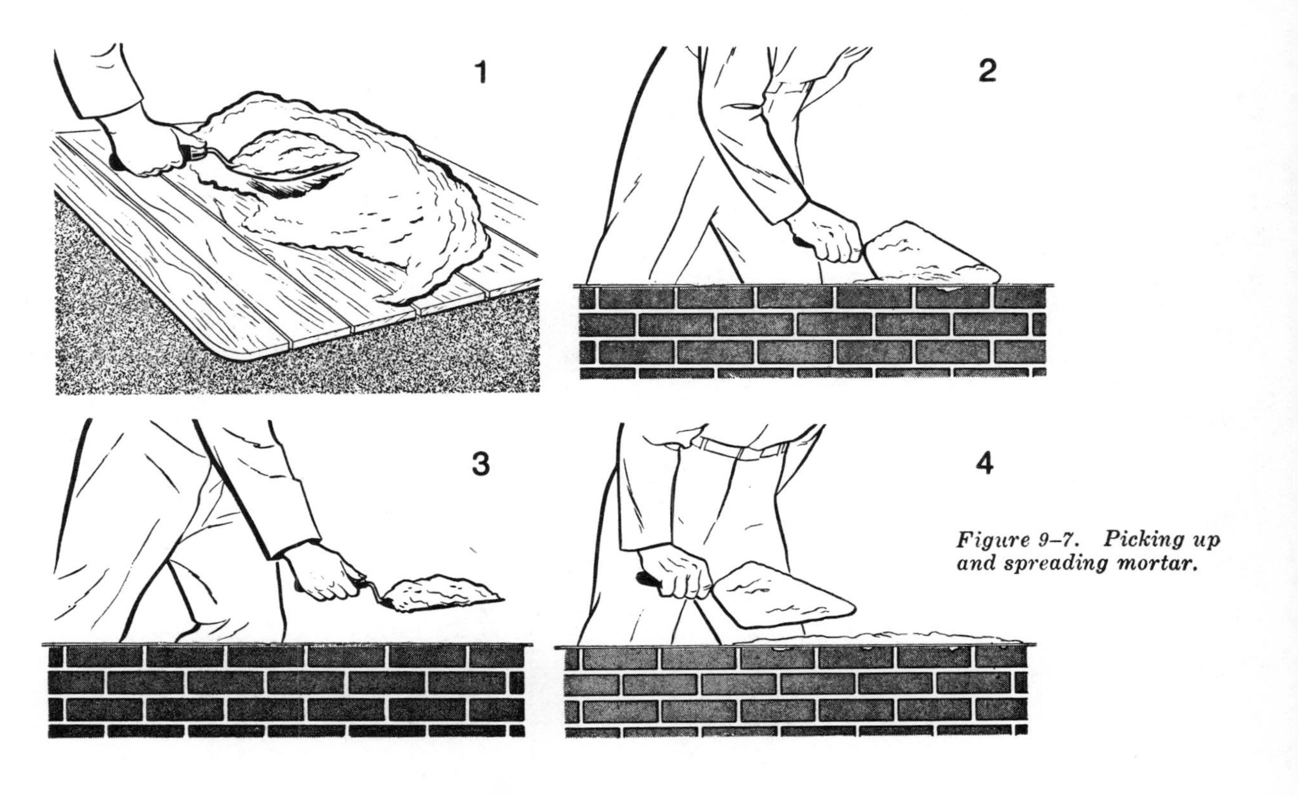

Figure 9–7. Picking up and spreading mortar.

Cutting Off Mortar. Mortar projecting beyond the wall line is cut off with the trowel edge (step 1, fig 9–8) and thrown back on the mortar board, but enough is retained to "butter" the left end of the first brick to be laid in the fresh mortar.

Bed Joint. With the mortar spread about 1 inch thick for the bed joint as shown in step 1, figure 9–8, a shallow furrow is made (step 2, figure 9–8) and the brick pushed into the mortar (step 3, figure 9–8). If the furrow is too deep, there will be a gap left between the mortar and the brick bedded in the mortar. This gap will reduce the resistance of the wall to water penetration. The mortar for a bed joint should not be spread out too far in advance of the laying. A distance of 4 or 5 bricks is advisable. Mortar that has been spread out too far will dry out before the brick is bedded in it. This results in a poor bond as can be seen by figure 9–9. The mortar must be soft and plastic so that the brick can be easily bedded in it.

Head Joint. The next step after the bed joint mortar has been spread is the laying of the brick. The brick to be laid is picked up as shown in figure 9–10 with the thumb on one side of the brick and the fingers on the other. As much mortar as will stick is placed on the end of the brick. The brick should then be pushed into place so that excess mortar squeezes out at the head joint and at the sides of the wall as indicated in figure 9–11. The head joint must be completely filled with mortar. This can only be done by placing plenty of mortar on the end of the brick. After the brick is bedded, the excess mortar is cut off and used for the next end joint. Surplus mortar should be thrown to the back of the mortar board for retempering if necessary. The proper position of the brick is determined by the use of a cord which can be seen in step 1, figure 9–14.

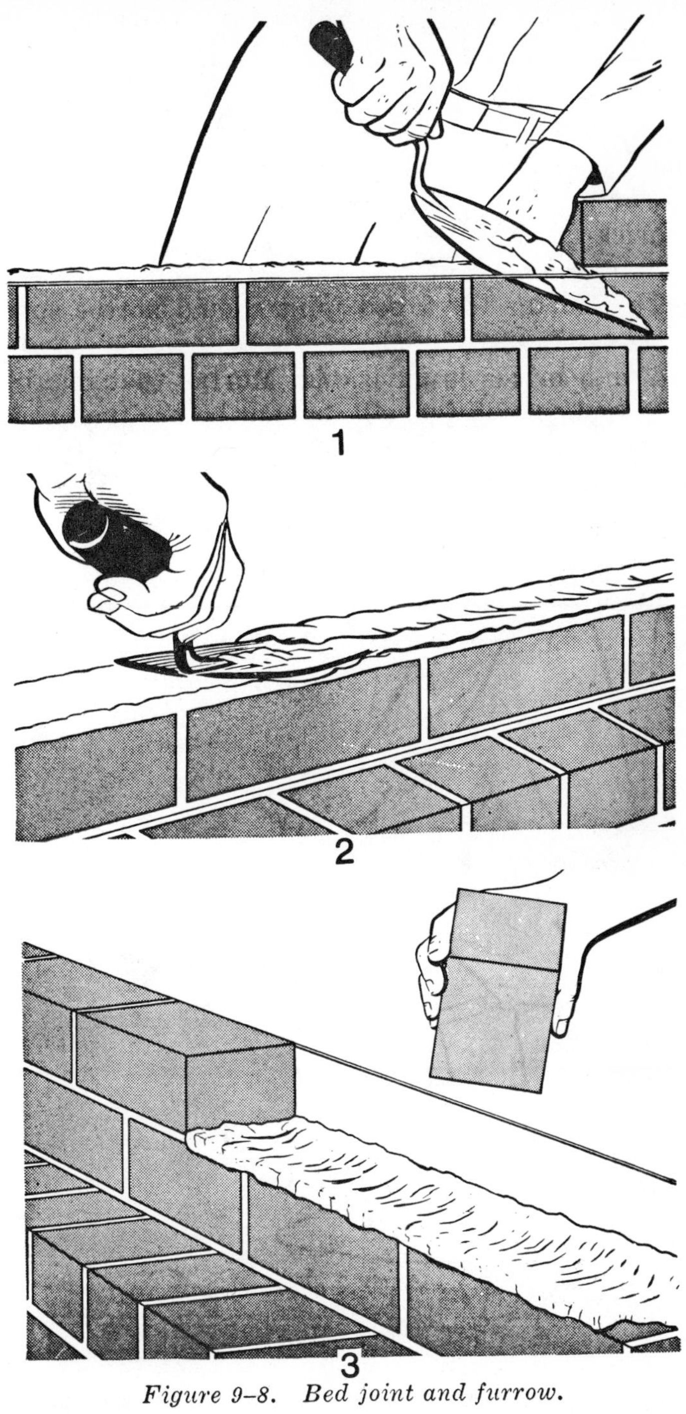

Figure 9–8. Bed joint and furrow.

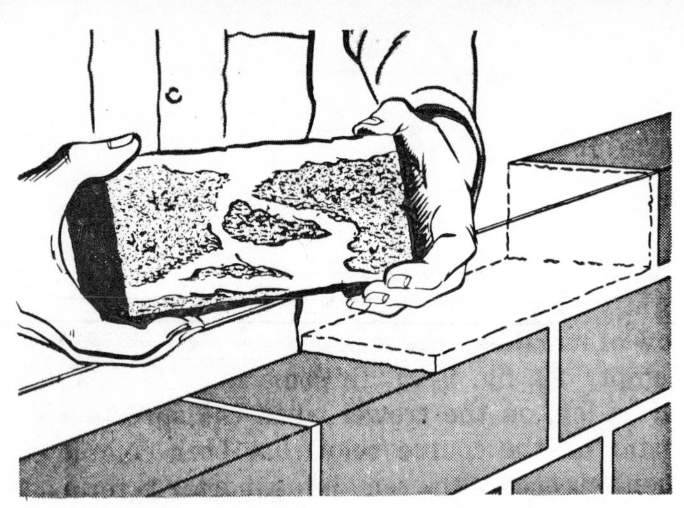

Figure 9–9. A poorly bonded brick.

Method of Inserting Brick in Wall. The method of inserting a brick in a space left in a wall is shown in figure 9–12. A thick bed of mortar is spread (step 1, fig. 9–12) and the brick

Figure 9–10. Proper way to hold a brick.

Figure 9–11. Head joint in a stretcher course.

shoved into this deep bed of mortar (step 2, fig. 9–12) until squeezes out at the top of the joint at the face tier, and at the header joint (step 3, fig. 9–12) so that the joints are full of mortar at every point.

Cross Joints in Header Courses. The position of a cross joint is illustrated in figure 9–13. These joints must be completely filled with mortar. The mortar for the bed joint should be spread several brick widths in advance. The mortar is spread over the entire side of the header brick before it is placed in the wall (step 1, fig. 9–13). The brick is then shoved into place so that the mortar is forced out at the top of the joint and the excess mortar cut off, as shown in step 2, figure 9–13.

Closure Joints in Header Courses. Figure 9–14 shows the method of laying a closure brick in a header course. Before laying the closure brick, plenty of mortar should be placed on the sides of

Figure 9–12. Laying inside brick

the brick already in place (step 1, fig. 9–14). Mortar should also be spread on both sides of the closure brick to a thickness of about 1 inch (step 2, fig. 9–14). The closure brick should then be laid in position without disturbing the brick already in place (step 3, fig. 9–14).

Closure Joints in Stretcher Courses. Before laying a closure brick for a stretcher course, the ends of the brick on each side of the opening to be filled with the closure brick should be well covered with mortar (step 1, fig. 9–15). Plenty of mortar should then be thrown on both ends of the closure brick (step 2, fig. 9–15) and the brick laid without

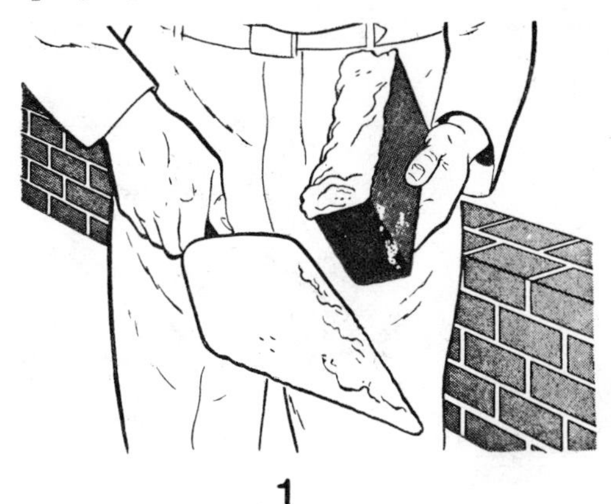

1

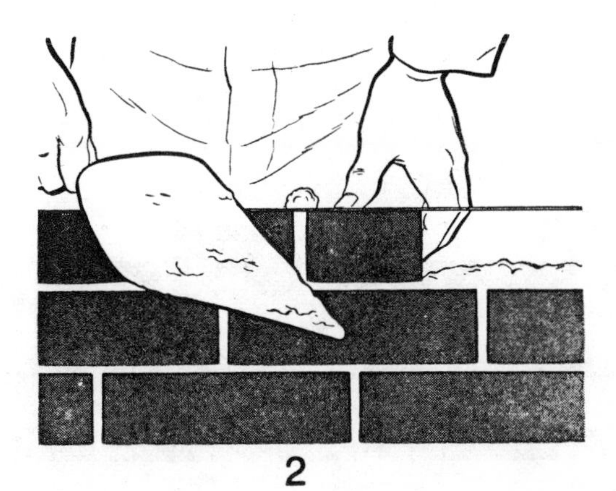

2

Figure 9–13. Making cross joints in header courses.

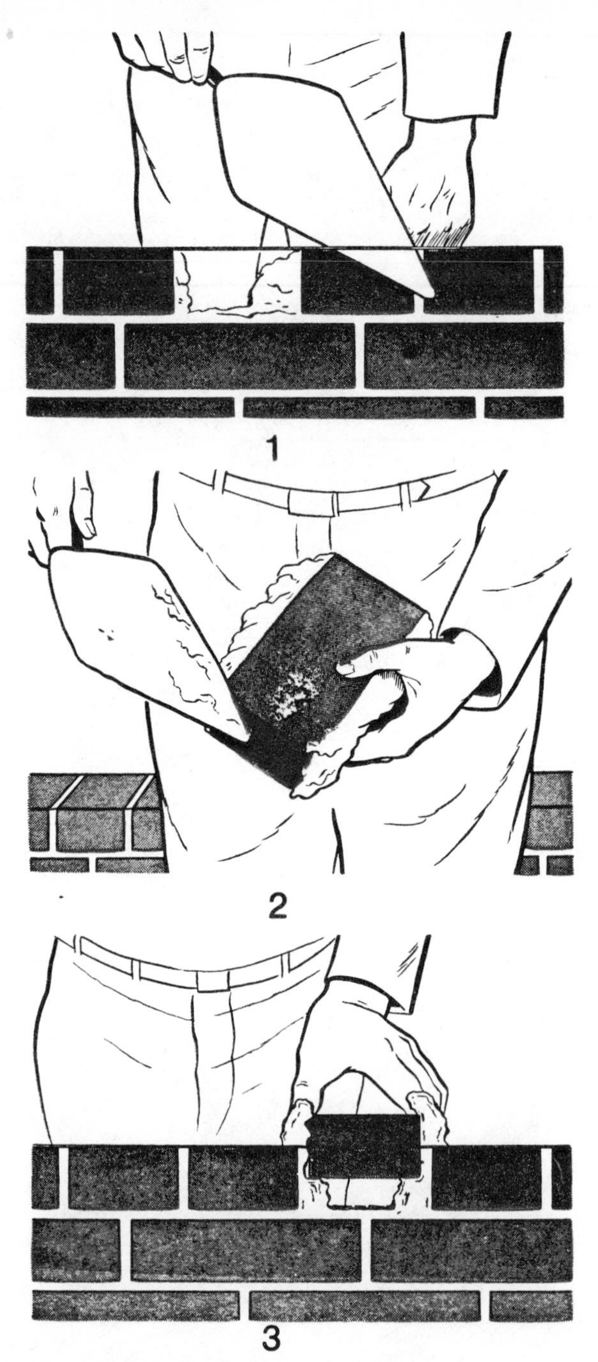

Figure 9–14. Making closure joints in header courses.

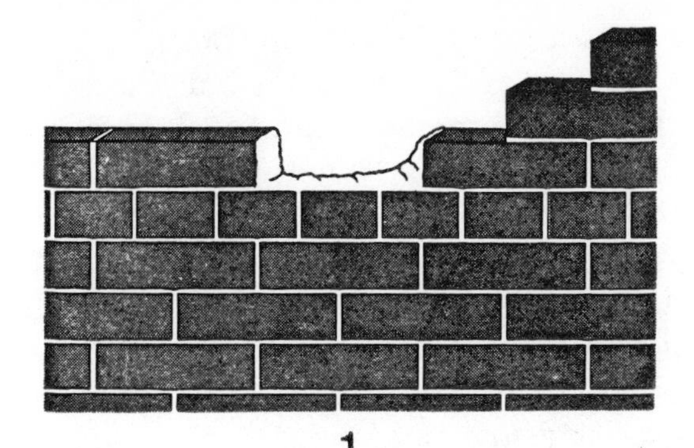

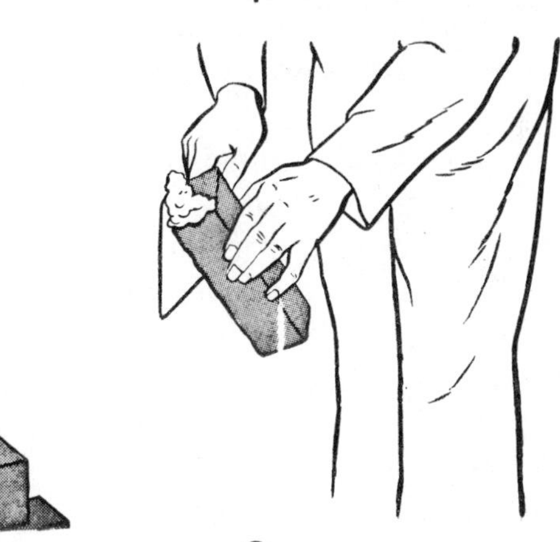

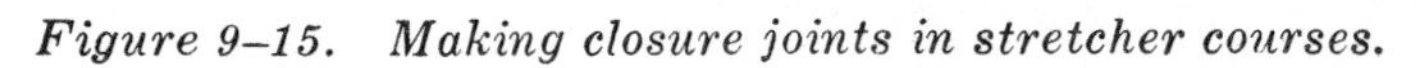

Figure 9–15. Making closure joints in stretcher courses.

disturbing those already in place (step 3, fig. 9–15). If any of the adjacent brick are disturbed they must be removed and relaid. Otherwise, cracks will form between the brick and mortar, allowing moisture into the wall.

Thickness of Mortar Joints. There is no hard and fast rule regarding the thickness of the mortar joint. Brick that are irregular in shape may require mortar joints up to ½-inch thick. All brick irregularities are taken up in the mortar joint. Mortar joints ¼ inch thick are the strongest and should be used when the bricks are regular enough to permit it.

Slushed Joints. Slushed joints are those made by depositing the mortar on the head joints in order that the mortar will run down between the brick to form a solid joint. *This should not be done.* Even when the space between the brick is completely filled, there is no way to compact the mortar against the faces of the brick and *a poor bond will result.*

Pointing. Filling exposed joints with mortar immediately after the wall has been laid is called pointing. Pointing is frequently necessary to fill holes and correct defective mortar joints. The pointing trowel is used for this purpose.

Cutting Brick

Cutting with Bolster or Brick Set. If a brick is to be cut to exact line the bolster or brick set should be used. When using these tools, the straight side of the cutting edge should face the part of the brick to be saved and also face the bricklayer. One blow of the hammer on the brick set should be enough to break the brick. Extremely hard brick will need to be cut roughly with the head of the hammer in such a way that there is enough brick left to be cut accurately with the brick set. See figure 9–16.

Figure 9–16. Cutting brick with a bolster.

Cutting with the Hammer. For normal cutting work, such as is required for making the closures and bats required around openings in walls and for the completion of corners, the brick hammer should be used. The first step is to cut a line all the way around the brick with light blows of the hammer head (fig. 9–17). When the line is complete, a sharp blow to one side of the cutting line will split the brick at the cutting line. Rough places are trimmed using the blade of the hammer, as shown in figure 9–17. The brick can be held in the hand while being cut.

Joint Finishes

Exterior Surfaces. Exterior surfaces of mortar joints are finished to make the brickwork more waterproof and to improve the appearance. There are several types of joint finishes, as shown in figure 9–18. The more important of these are discussed below. When joints are cut flush with the brick and not finished, cracks are immediately apparent between the brick and the mortar. Although these cracks are not deep, they are unde-

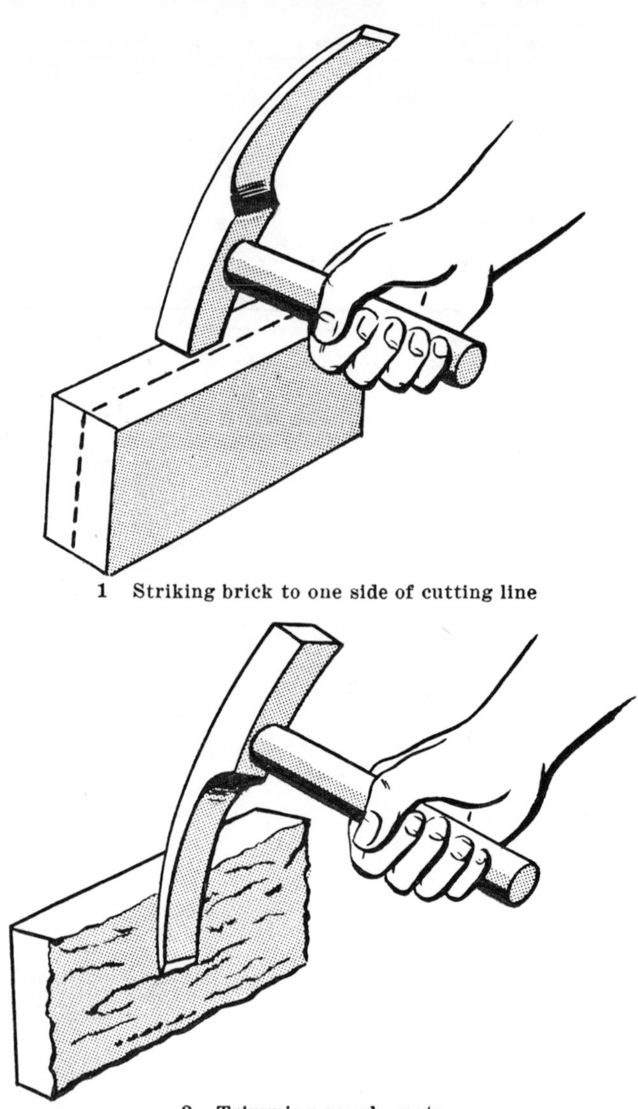

1 Striking brick to one side of cutting line

2 Trimming rough spots

Figure 9–17. Cutting brick with a hammer.

sirable and can be eliminated by finishing or tooling the joint. In every case, the mortar joint should be finished before the mortar has hardened to any appreciable extent. The jointing tool is shown in figure 7–1.

Concave Joint. The best joint from the standpoint of weather-tightness is the concave joint.

This joint is made with a special tool after the excess mortar has been removed with the trowel. The tool should be slightly larger than the joint. Force is used to press the mortar tight against the brick on both sides of the mortar joint.

Flush Joint. The flush joint (fig. 9–18) is made by keeping the trowel almost parallel to the face of the wall while drawing the point of the trowel along the joint.

Weather Joint. A weather joint sheds water more easily from the surface of the wall and is

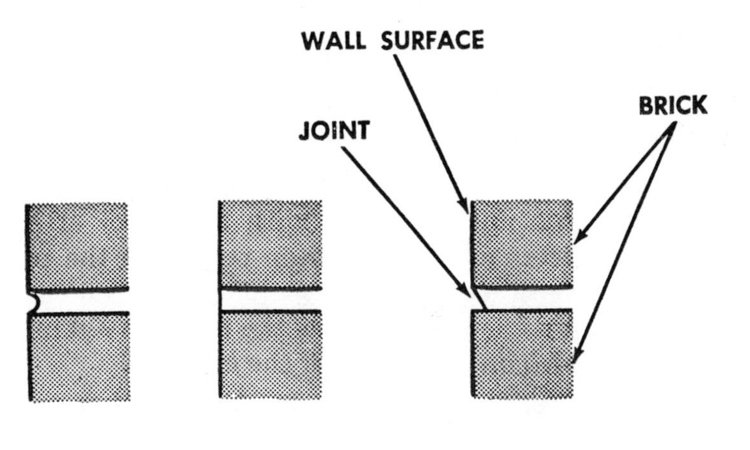

Figure 9–18. Joint finishes.

formed by pushing downward on the mortar with the top edge of the trowel.

BRICK CONSTRUCTION

The bricktender mixes mortar, carries brick and mortar to the bricklayer, and keeps him supplied with these materials at all times. He fills the mortar board and places it in a position convenient for the bricklayer. He assists in the layout and, at times, such as during rapid backup bricklaying, he may lay out brick in a line on an adjacent course so that the bricklayer needs to move each brick only a few inches in laying backup work.

Wetting brick is also the duty of the bricktender. This is done when bricks are laid in warm weather. There are four reasons for wetting brick just before they are laid:

There will be a better bond between the brick and the mortar.

The water will wash dust and dirt from the surface of the brick. Mortar adheres better to a clean brick.

If the surface of the brick is wet, the mortar spreads more evenly under it.

A dry brick may absorb water from the mortar rapidly. This is particularly bad when mortar containing portland cement is used. In order for cement to harden properly, sufficient moisture must be present to complete the hydration of the cement. If the brick robs the mortar of too much water, there will not be enough left to hydrate the cement properly.

The bricklayer does the actual laying of the brick. It is his responsibility to lay out the job so that the finished masonry will be properly done. In construction involving walls, he must see that the walls are plumb and the courses level.

Footings

Wall Footing.

A footing is required under a wall when the bearing capacity of the supporting soil is not sufficient to withstand the wall load without a further means of redistribution. The footing must be wider than the thicknes of the wall, as illustrated in figure 9–19. The required footing width and thickness for walls of considerable height or for walls that are to carry a heavy load should be determined by a qualified engineer. Every footing should be below the frost line in order to prevent heaving and settlement of the foundation. For the usual one-story building with an 8-inch-thick wall, a footing 16 inches wide and approximately 8 inches thick as given in figure 9–19 is usually enough. Although brickwork footings are satisfactory, footings are normally concrete, leveled on top to receive the brick or stone foundation wall.

As soon as the subgrade is prepared, the bricklayer should place a bed of mortar about 1 inch thick on the subgrade to take up all irregularities. The first course of the foundation is laid on this bed of mortar. The other courses are then laid on this first course (fig. 9–19).

Column Footing. A column footing for a 12- by 16-inch brick column is shown in figure 9–20. The construction method for this footing is the same as for the wall footing.

Eight-Inch Common Bond Brick Wall

Laying Out the Wall. For a wall of given length, the bricklayer makes a slight adjustment in the width of head joints so that some number of brick, or some number including one half-brick, will just make up the length. The bricklayer first lays the brick on the foundation without mortar

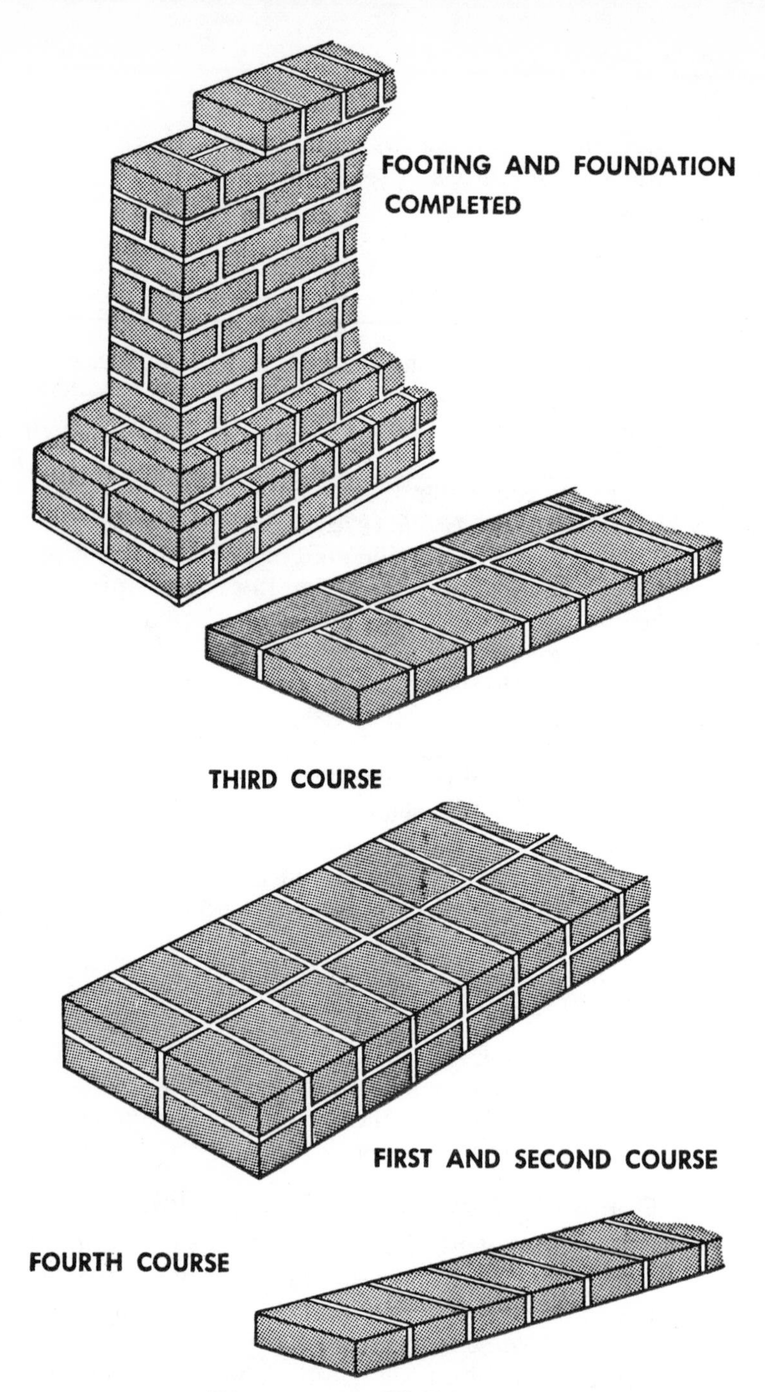

Figure 9–19. Wall footing.

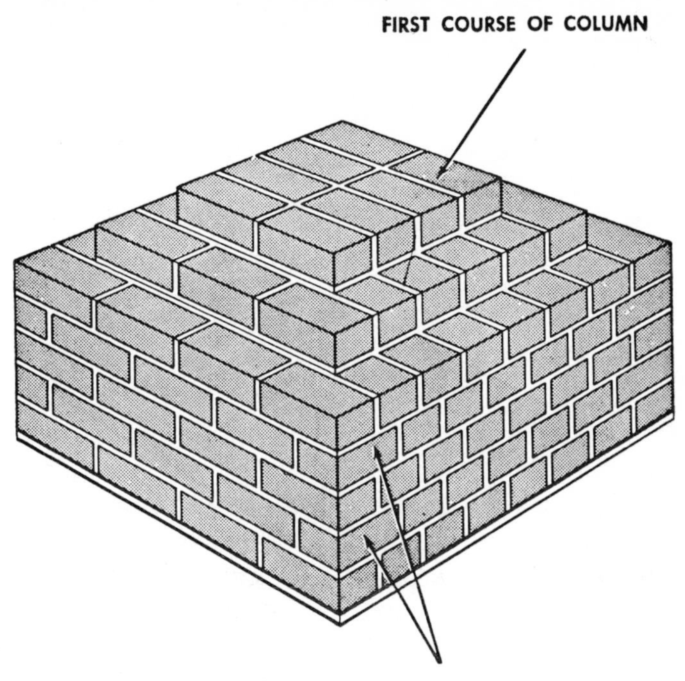

Figure 9–20. Column footing.

(fig. 9–21). The distance between the bricks is equal to the thickness of the head mortar joints. Tables 9–2, 9–3, and 9–4 give the number of courses and horizontal joints required for a given wall height.

Laying Corner Leads. The corners are erected first. This is called "laying of leads." The bricklayer will use these leads as a guide in laying the remainder of the wall. Step 3, figure 9–23 shows a corner lead laid up and racked so that the center portion of the wall can be bonded into the corner lead. Only the face tier is laid in these corners and the corner lead is built up above the rest of the wall a distance of six or seven courses or to the height of the next header course above. Normally the first course is a header course.

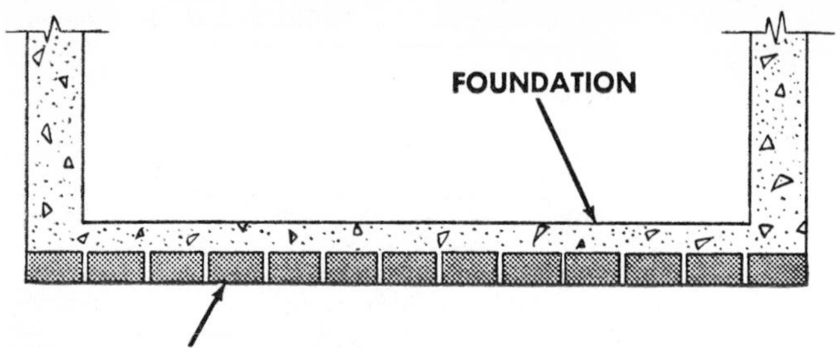

Figure 9–21. Determination of vertical brick joints and number of brick in one course.

The first step in laying a corner lead is shown in first step, figure 9–22. Two three-quarter closures are cut and a 1-inch thick mortar bed is laid on the foundation. The three-quarter closure marked by *a* in second step, figure 9–22 is pressed down into the mortar bed until the bed joint becomes ½ inch thick. Next, mortar is placed on the end of three-quarter closure *b* and a head joint is formed as described in paragraph 9–8. The head joint between the two three-quarter closures should be ½ inch thick also. Excess mortar that has been squeezed out of the joints is cut off. The level of the two three-quarter closures should now be checked by means of the plumb rule placed in the positions indicated by the heavy dashed lines in second step, figure 9–22. The edges of both these closure bricks must be even with the outside face of the foundation.

Next mortar is spread on the side of brick *c* and it is laid as shown in third step, figure 9–22. Its level is checked using the plumb rule in the position given in third step, figure 9–22. Its end must also be even with the outside face of the foundation. Brick *d* is laid and its level and position checked. When brick *d* is in the proper position, the quarter closures *e* and *f* should be cut and placed according to the recommended procedure

Table 9-2. Height of Courses: 2¼-Inch Brick, ⅜-Inch Joint

Courses	Height	Courses	Height	Courses	Height	Courses	Height	Courses	Height
1	0′ 2⅝″	21	4′ 7⅛″	41	8′ 11⅝″	61	13′ 4⅛″	81	17′ 8⅝″
2	0′ 5¼″	22	4′ 9¾″	42	9′ 2¼″	62	13′ 6¾″	82	17′ 11¼″
3	0′ 7⅞″	23	5′ 0⅜″	43	9′ 4⅞″	63	13′ 9⅜″	83	18′ 1⅞″
4	0′ 10½″	24	5′ 3″	44	9′ 7½″	64	14′ 0″	84	18′ 4½″
5	1′ 1⅛″	25	5′ 5⅝″	45	9′ 10⅛″	65	14′ 2⅝″	85	18′ 7⅛″
6	1′ 3¾″	26	5′ 8¼″	46	10′ 0¾″	66	14′ 5¼″	86	18′ 9¾″
7	1′ 6⅜″	27	5′ 10⅞″	47	10′ 3⅜″	67	14′ 7⅞″	87	19′ 0⅜″
8	1′ 9″	28	6′ 1½″	48	10′ 6″	68	14′ 10½″	88	19′ 3″
9	1′ 11⅝″	29	6′ 4⅛″	49	10′ 8⅝″	69	15′ 1⅛″	89	19′ 5⅝″
10	2′ 2¼″	30	6′ 6¾″	50	10′ 11¼″	70	15′ 3¾″	90	19′ 8¼″
11	2′ 4⅞″	31	6′ 9⅜″	51	11′ 1⅞″	71	15′ 6⅜″	91	19′ 10⅞″
12	2′ 7½″	32	7′ 0″	52	11′ 4½″	72	15′ 9″	92	20′ 1½″
13	2′ 10⅛″	33	7′ 2⅝″	53	11′ 7⅛″	73	15′ 11⅝″	93	20′ 4⅛″
14	3′ 0¾″	34	7′ 5¼″	54	11′ 9¾″	74	16′ 2¼″	94	20′ 6¾″
15	3′ 3⅜″	35	7′ 7⅞″	55	12′ 0⅜″	75	16′ 4⅞″	95	20′ 9⅜″
16	3′ 6″	36	7′ 10½″	56	12′ 3″	76	16′ 7½″	96	21′ 0″
17	3′ 8⅝″	37	8′ 1⅛″	57	12′ 5⅝″	77	16′ 10⅛″	97	21′ 2⅝″
18	3′ 11¼	38	8′ 3¾″	58	12′ 8¼″	78	17′ 0¾″	98	21′ 5¼″
19	4′ 1⅞″	39	8′ 6⅜″	59	12′ 10⅞″	79	17′ 3⅜″	99	21′ 7⅞″
20	4′ 4½″	40	8′ 9″	60	13′ 1½″	80	17′ 6″	100	21′ 10½″

Table 9–3. Height of Courses: 2¼-Inch Brick, ½-Inch Joint

Courses	Height	Courses	Height	Courses	Height	Courses	Height	Courses	Height
1	0′ 2¾″	21	4′ 9¾″	41	9′ 4¾″	61	13′ 11¾″	81	18′ 6¾″
2	0′ 5½″	22	5′ 0½″	42	9′ 7½″	62	14′ 2½″	82	18′ 9½″
3	0′ 8¼″	23	5′ 3¼″	43	9′ 10¼″	63	14′ 5¼″	83	19′ 0¼″
4	0′ 11″	24	5′ 6″	44	10′ 1″	64	14′ 8″	84	19′ 3″
5	1′ 1¾″	25	5′ 8¾″	45	10′ 3¾″	65	14′ 10¾″	85	19′ 5¾″
6	1′ 4½″	26	5′ 11½″	46	10′ 6½″	66	15′ 1½″	86	19′ 8½″
7	1′ 7¼″	27	6′ 2¼″	47	10′ 9¼″	67	15′ 4¼″	87	19′ 11¼″
8	1′ 10″	28	6′ 5″	48	11′ 0″	68	15′ 7″	88	20′ 2″
9	2′ 0¾″	29	6′ 7¾″	49	11′ 2¾″	69	15′ 9¾″	89	20′ 4¾″
10	2′ 3½″	30	6′ 10½″	50	11′ 5½″	70	16′ 0½″	90	20′ 7½″
11	2′ 6¼″	31	7′ 1¼″	51	11′ 8¼″	71	16′ 3¼″	91	20′ 10¼″
12	2′ 9″	32	7′ 4″	52	11′ 11″	72	16′ 6″	92	21′ 1″
13	2′ 11¾′	33	7′ 6¾″	53	12′ 1¾″	73	16′ 8¾″	93	21′ 3¾″
14	3′ 2½″	34	7′ 9½″	54	12′ 4½″	74	16′ 11½″	94	21′ 6½″
15	3′ 5¼″	35	8′ 0¼″	55	12′ 7¼″	75	17′ 2¼″	95	21′ 9¼″
16	3′ 8″	36	8′ 3″	56	12′ 10″	76	17′ 5″	96	22′ 0″
17	3′ 10¾″	37	8′ 5¾″	57	13′ 0¾″	77	17′ 7¾″	97	22′ 2¾″
18	4′ 1½″	38	8′ 8½″	58	13′ 3½″	78	17′ 10½″	98	22′ 5½″
19	4′ 4¼″	39	8′ 11¼″	59	13′ 6¼″	79	18′ 1¼″	99	22′ 8¼″
20	4′ 7″	40	9′ 2″	60	13′ 9″	80	18′ 4″	100	22′ 11″

Table 9–4. Height of Courses: 2¼-Inch Brick, ⅝-Inch Joint

Courses	Height	Courses	Height	Courses	Height	Courses	Height	Courses	Height
1	0′ 2⅞″	21	5′ 0⅜″	41	9′ 9⅞″	61	14′ 7⅜″	81	19′ 4⅞″
2	0′ 5¾″	22	5′ 3¼″	42	10′ 0¾″	62	14′ 10¼″	82	19′ 7¾″
3	0′ 8⅝″	23	5′ 6⅛″	43	10′ 3⅝″	63	15′ 1⅛″	83	19′ 10⅝″
4	0′ 11½″	24	5′ 9″	44	10′ 6½″	64	15′ 4″	84	20′ 1½″
5	1′ 2⅜″	25	5′ 11⅞″	45	10′ 9⅜″	65	15′ 6⅞″	85	20′ 4⅜″
6	1′ 5¼″	26	6′ 2¾″	46	11′ 0¼″	66	15′ 9¾″	86	20′ 7¼″
7	1′ 8⅛″	27	6′ 5⅝″	47	11′ 3⅛″	67	16′ 0⅝″	87	20′ 10⅛″
8	1′ 11″	28	6′ 8½″	48	11′ 6″	68	16′ 3½″	88	21′ 1″
9	2′ 1⅞″	29	6′ 11⅜″	49	11′ 8⅞″	69	16′ 6⅜″	89	21′ 3⅞″
10	2′ 4¾″	30	7′ 2¼″	50	11′ 11¾″	70	16′ 9¼″	90	21′ 6¾″
11	2′ 7⅝″	31	7′ 5⅛″	51	12′ 2⅝″	71	17′ 0⅛″	91	21′ 9⅝″
12	2′ 10½″	32	7′ 8″	52	12′ 5½″	72	17′ 3″	92	22′ 0½″
13	3′ 1⅜″	33	7′ 10⅞″	53	12′ 8⅜″	73	17′ 5⅞″	93	22′ 3⅜″
14	3′ 4¼″	34	8′ 1¾″	54	12′ 11¼″	74	17′ 8¾″	94	22′ 6¼″
15	3′ 7⅛″	35	8′ 4⅝″	55	13′ 2⅛″	75	17′ 11⅝″	95	22′ 9⅛″
16	3′ 10″	36	8′ 7½″	56	13′ 5″	76	18′ 2½″	96	23′ 0″
17	4′ 0⅞″	37	8′ 10⅜″	57	13′ 7⅛″	77	18′ 5⅜	97	23′ 2⅞″
18	4′ 3¾″	38	9′ 1¼″	58	13′ 10¾″	78	18′ 8¼″	98	23′ 5¾″
19	4′ 6⅝″	39	9′ 4⅛″	59	14′ 1⅝″	79	18′ 11⅛″	99	23′ 8⅝″
20	4′ 9½″	40	9′ 7″	60	14′ 4½″	80	19′ 2″	100	23′ 11½″

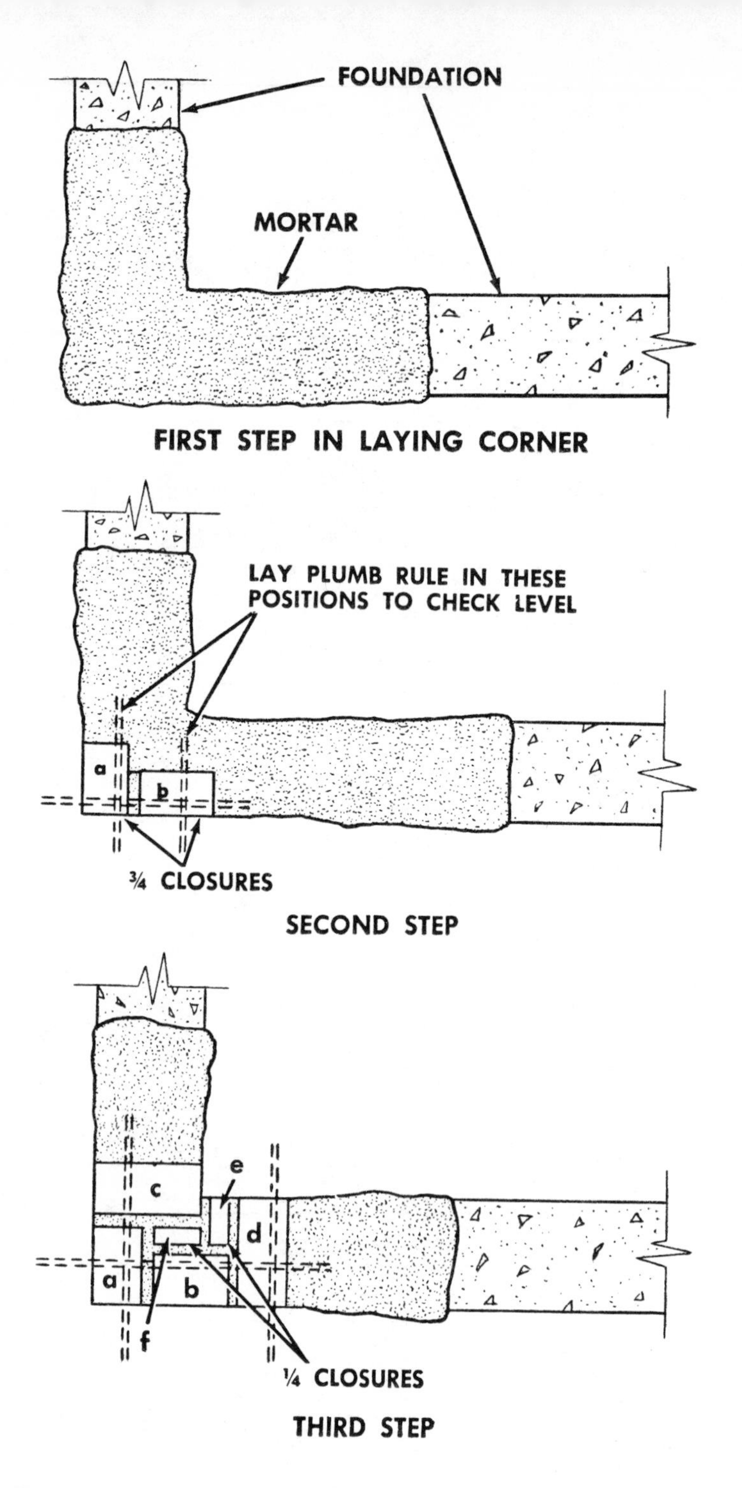
FOUNDATION
MORTAR
FIRST STEP IN LAYING CORNER
LAY PLUMB RULE IN THESE
POSITIONS TO CHECK LEVEL
a
b
¾ CLOSURES
SECOND STEP
c
e
d
a
b
f
¼ CLOSURES
THIRD STEP

Figure 9–22. First course of corner lead for 8-inch common bond brick wall.

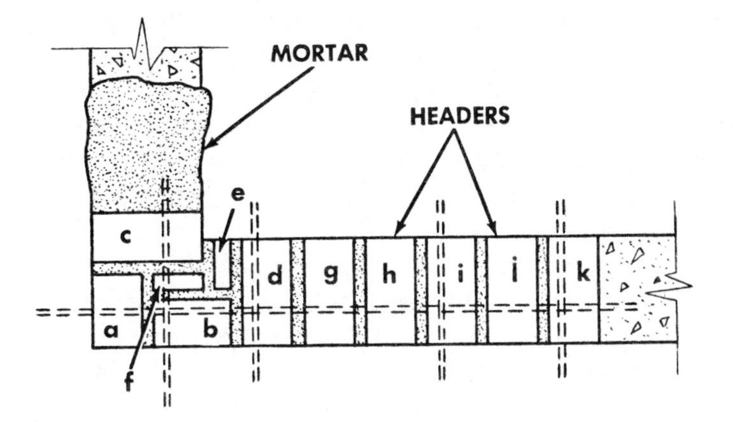

FOURTH STEP

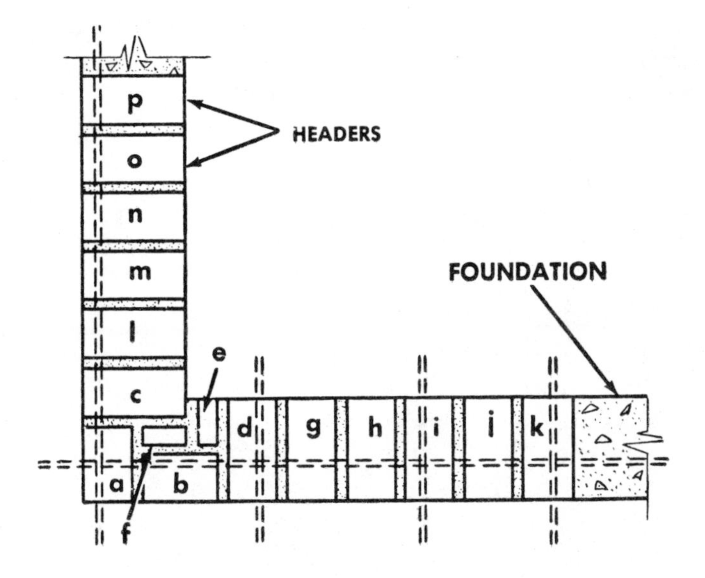

FIFTH STEP

described for laying closure brick. All excess mortar should be removed and the tops of these quarter closures checked to see that they are at the same level as the tops of surrounding brick.

Brick *g* (fourth step, fig. 9–22) is now shoved into position after mortar has been spread on its face. Excess mortar should be removed. Bricks *h, i, j,* and *k* are laid in the same manner. The level of the brick is checked by placing the plumb rule in the several positions indicated in fourth step, figure 9–22. All brick ends must be flush with the surface of the foundation. Bricks *l, m, n, o,* and *p* are then laid in the same manner. The number of leader bricks that must be laid in the first course of the corner lead can be determined from fifth step, figure 9–22. It will be noted that six header bricks are required on each side of the three-quarter closures a and b.

The second course, a stretcher course, is now laid. Procedure is shown in step 1, figure 9–23. A 1-inch thick layer of mortar should be spread over the first course and a shallow furrow made in the mortar bed. Brick *a* (step 2, figure 9–23) is then laid in the mortar bed and shoved down until the mortar joint is ½-inch thick. Brick *b* may now be shoved into place after mortar has been spread on its end. Excess mortar is removed and the joint checked for thickness. Bricks *c, d, e, f,* and *g* are laid in the same manner and checked to make them level and plumb. The level is checked by placing the plumb rule in the position indicated in step 2, figure 9–23. The bricks are plumbed by using the plumb rule in a vertical position as shown in figure 9–24. This should be done in several places. As may be determined from 3, figure 9–23, seven bricks are required for the second course. The remaining bricks in the corner lead are laid in the manner described for the bricks in the second course.

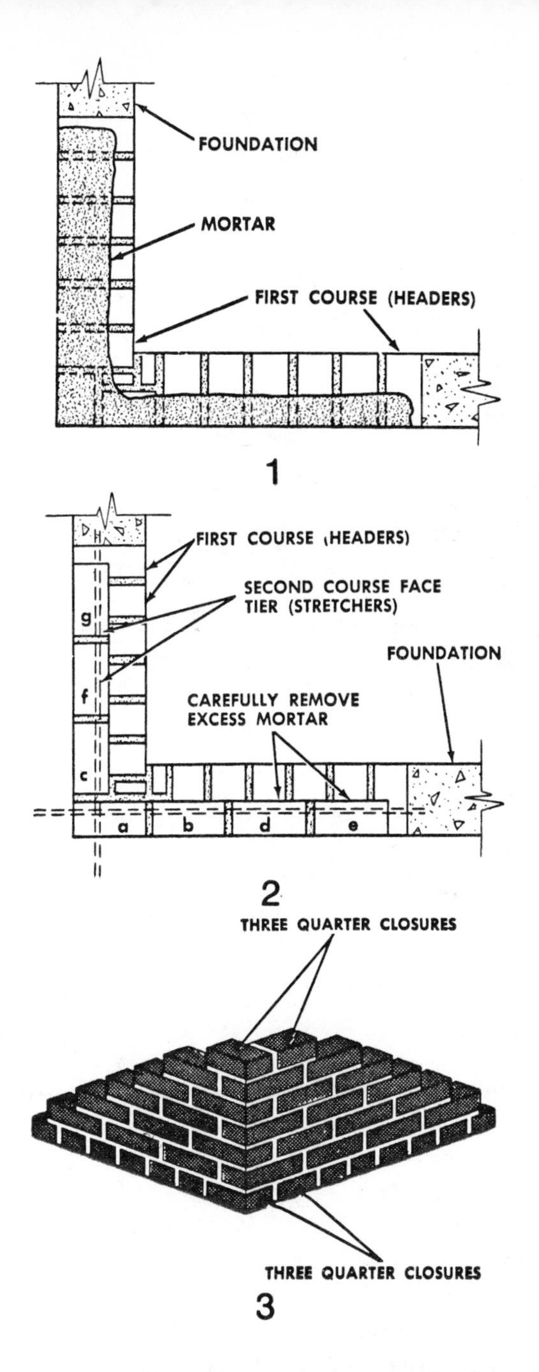

Figure 9-23. Second course of corner lead for 8-inch common bond brick wall.

Figure 9–24. Plumbing a corner.

Since the portion of the wall between the leads is laid using the leads as a guide, the level of the courses in the lead must be checked continually, and after the first few courses the lead is plumbed. If the brickwork is not plumb, bricks must be moved in or out until the lead is accurately plumb. It is not good practice to move brick much once they are laid in mortar; therefore, care is taken to place the brick accurately at the start. Before the mortar has set, the joints are tooled or finished.

A corner lead at the opposite end of the wall is built in the same manner. It is essential that the level of the tops of corresponding courses be the same in each lead; that is, the top of the second course in one corner lead must be at the same height above the foundation as the second course in the other corner lead. A long 2-by 2-inch pole can be used to mark off the heights of the different courses above the foundation. This pole

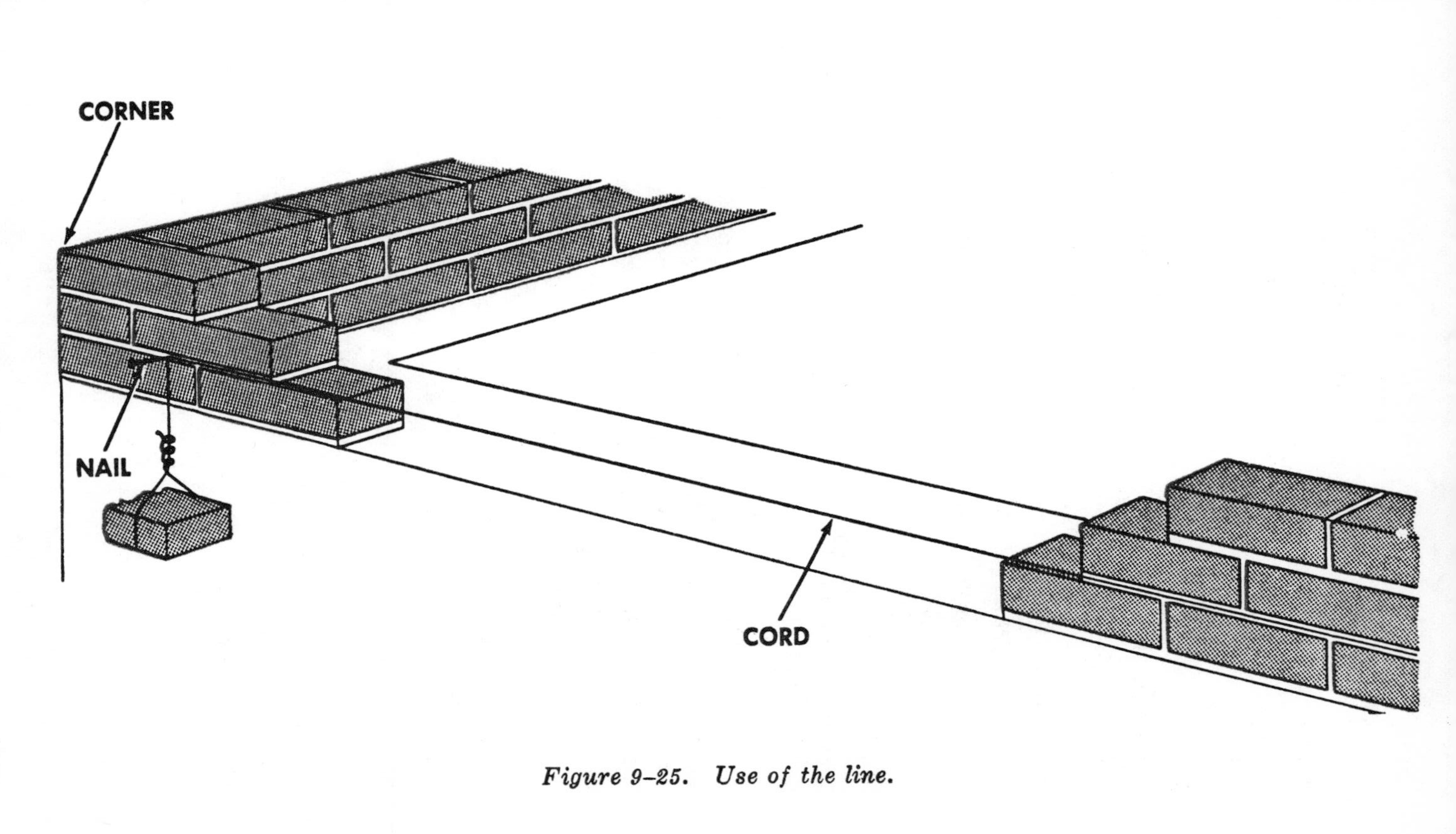

Figure 9–25. Use of the line.

can be used to check the course height in the corner leads. The laying of leads is closely supervised and only skilled men are employed in this work.

Laying the Face Tier Between the Corner Leads.

With the corner leads at each end of the wall completed, the face tier of brick for the wall between the leads is laid. It is necessary to use a line, as shown in figure 9–25. Nails are driven into the top of the mortar joint as shown in figure 9–25. The line is hung over the nails and pulled taut by means of weights attached to each end. The line is positioned $1/16$ inch outside the wall face level with the top of the brick.

With the line in place, the first or header course is laid in place between the two corner leads, as described in paragraph 9–8. The brick is shoved into position so that its top edge is $1/16$ inch behind the line. Do not crowd the line. If the corner leads are accurately built, the entire wall will be level and plumb. It is not necessary to use the level on the section of the wall between the leads; however, it is advisable to check it with the level at several points. For the next course, the line is moved to the top of the next mortar joint. The brick in the stretcher course should be laid as described in paragraph 9–8. Finish the face joints before the mortar hardens.

When the face tier of brick for the wall between the leads has been laid up to but not including the second header course, normally six courses, the backup tier is laid. Procedure for laying backup brick has already been described. The backup brick for the corner leads are laid first and the remaining brick afterwards (fig. 9–26). The line need not be used for the backup brick in an 8-inch wall. When the backup brick have been laid up to the height of the second header course, the second header course is laid.

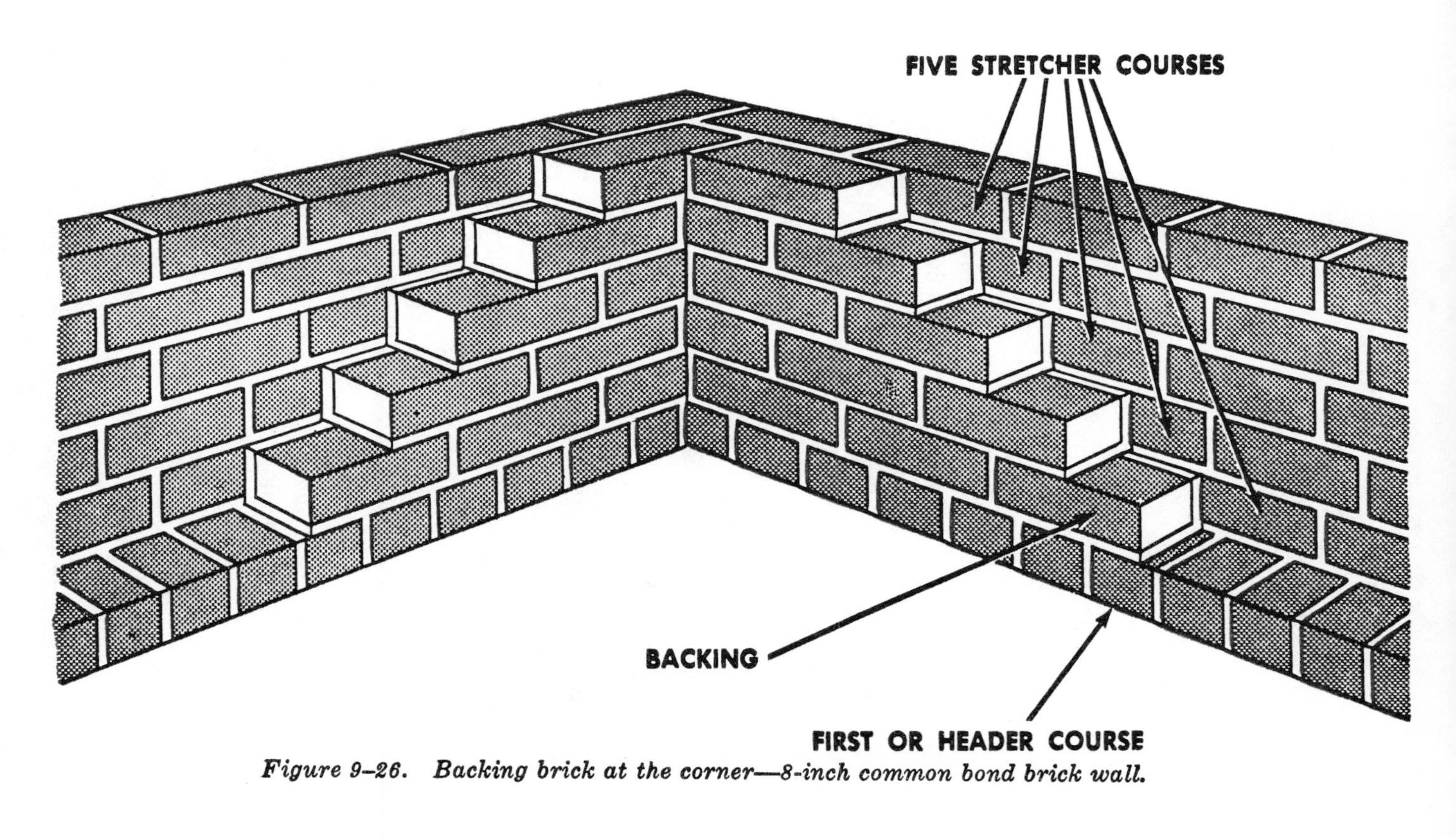

Figure 9–26. Backing brick at the corner—8-inch common bond brick wall.

The wall for the entire building is built up to a height including the second header course at which time corner leads are continued six more courses. The wall between the leads is constructed as before and the entire procedure repeated until the wall has been completed to the required height.

Twelve-Inch Common Bond Brick Wall

The 12-inch-thick common bond brick wall is laid out as shown in 3, figure 9–27. Note that the construction is similar to that for the 8-inch wall with the exception that a third tier of brick is used. The header course is laid (1, fig. 9–27) first and the corner leads built. Two tiers of backing brick are required instead of one. The second course is shown in 2, figure 9–27 and the third course in 3, figure 9–27. Two header courses are required and they overlap as shown in 1, figure 9–27. A line should be used for the inside tier of backing brick for a 12-inch wall.

Protection of Brickwork and Use of a Trig

Protection the Brick. The tops of all brick walls should be protected each night from rain damage by placing boards or tarpaulins on top of the wall and setting loose bricks on them.

Use of a Trig. When a line is stretched on a long wall, a trig is used to prevent sagging and to keep it from being blown in or out from the face of the wall by the wind. The trig consists of a short piece of line looped around the main line and fastened to the top edge of a brick that has been previously laid in proper position. A lead between the corner leads must be erected in order to place the trig brick in its proper location.

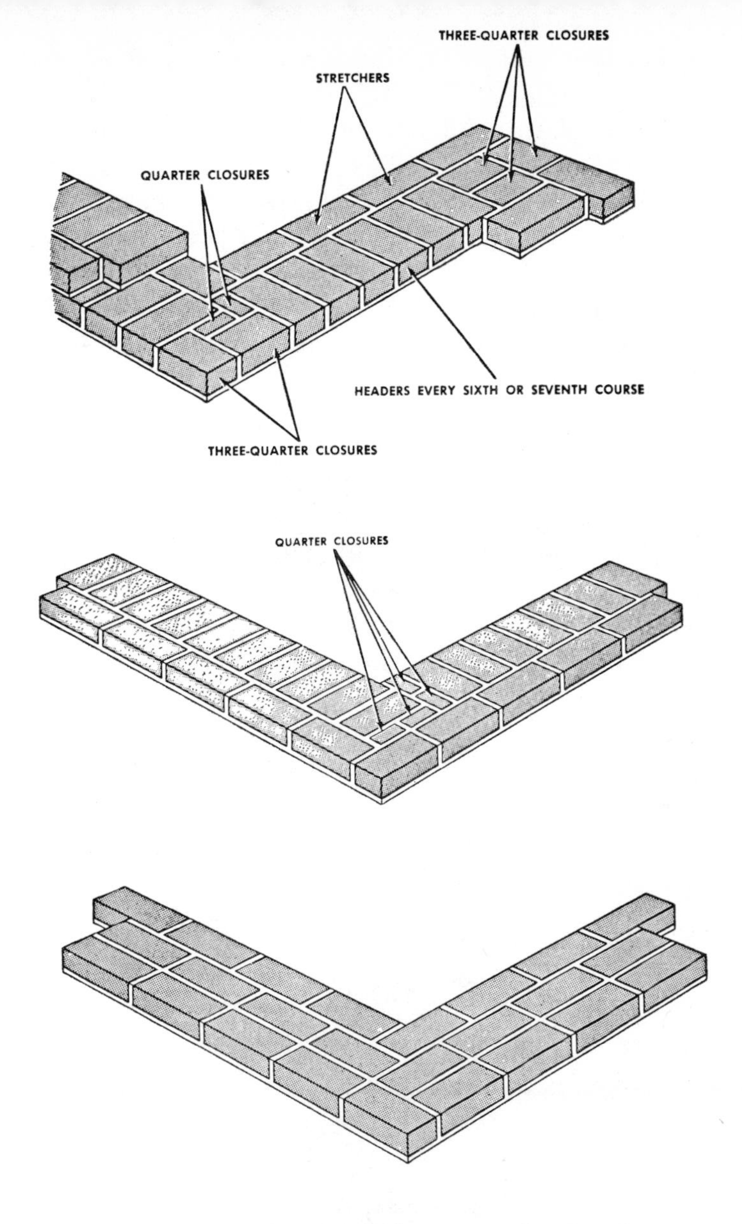

Figure 9–27. Twelve-inch common bond wall.

Window and Door Openings

If windows are to be installed in the wall, openings are left for them as the bricklaying proceeds. The height to the top of one full course should be exactly the height of the window sill. When the distance from the foundation to the bottom of the window sill is known, the bricklayer can determine how many courses are required to bring the wall up to that height. If the sill is to be 4 feet $4\frac{1}{4}$ inches above the foundation and $\frac{1}{2}$-inch mortar joints are to be used, 19 courses will be required. (Each brick plus one mortar joint is $2\frac{1}{4} + \frac{1}{2} = 2\frac{3}{4}$ inches. One course is thus $2\frac{3}{4}$ inches high. Four feet $4\frac{1}{4}$ inches divided by $2\frac{3}{4}$ is 19, the number of courses required.)

With the brick laid up to sill height, the rowlock sill course is laid as shown in figure 9–28. The rowlock course is pitched downward. The slope is away from the window and the rowlock course normally takes up a vertical space equal to two courses of brick. The exterior surface of the joints between the brick in the rowlock course must be carefully finish to make them watertight.

The window frame is placed on the rowlock sill as soon as the mortar has set. The window

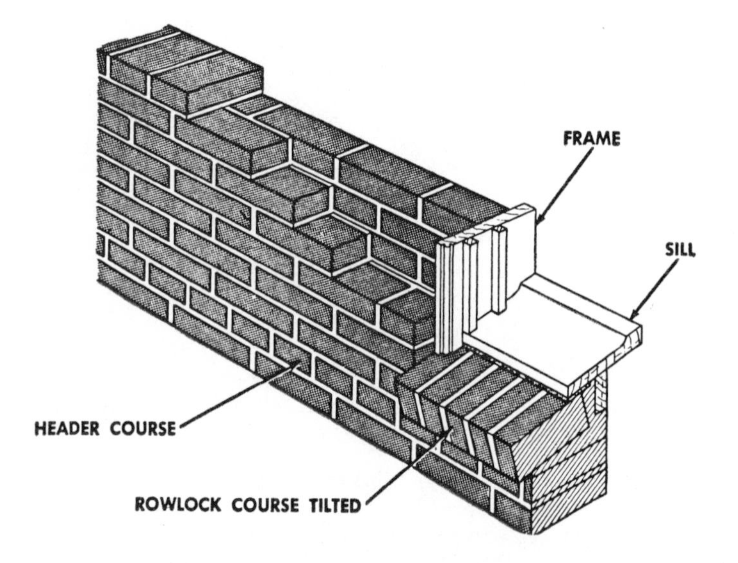

Figure 9–28. Construction of a window opening.

frame must be temporarily braced until the brickwork has been laid up to about one-third the height of the window frame. These braces are not removed for several days in order that the wall above the window frame will set properly. Now the bricklayer lays up the brick in the rest of the wall in such a way that the top of the brick in the course at the level of the top of the window frame is not more than 1/4 inch above the frame. To do this, he marks on the window frame with a pencil the top of each course. If the top course does not come to the proper level, he changes the thickness of the joints slightly until the top course is at the proper level. The corner leads should be laid up after the height of each course at the window is determined.

The mortar joint thickness for the corner leads is made the same as that determined at the window opening. With the corner leads erected, the line is installed as already described and is stretched across the window opening. The brick can now be laid in the rest of the wall. If the window openings have been planned properly, the brick in the face tier can be laid with a minimum of brick cutting.

Placing Lintel Over Window Opening. Lintels are placed above windows and doors to carry the weight of the wall above them. They rest on the brick course that is level or approximately level with the frame head, and are firmly bedded in mortar at the sides. Any space between the window frame and the lintel is closed with blocking and weatherstripped with bituminous materials. The wall is then continued above the window after the lintel is placed.

Door Openings. The same procedure can be used for laying brick around a door opening as was used for laying brick around a window opening, including placement of the lintel. The arrangement at a door opening is given in figure 9-29. Pieces of wood cut to the size of a half closure are laid in mortar as brick to provide for anchoring the door frame by means of screws or nails. These wood blocks are placed at several points along the top and sides of the door opening to allow for plumbing the frame.

Lintels

The brickwork above openings in walls must be supported by lintels. Lintels can be made of steel, precast reinforced concrete beams, or wood. The use of wood should be avoided as much as possible. If reinforced brick masonry is employed, the brick above the wall opening can be supported

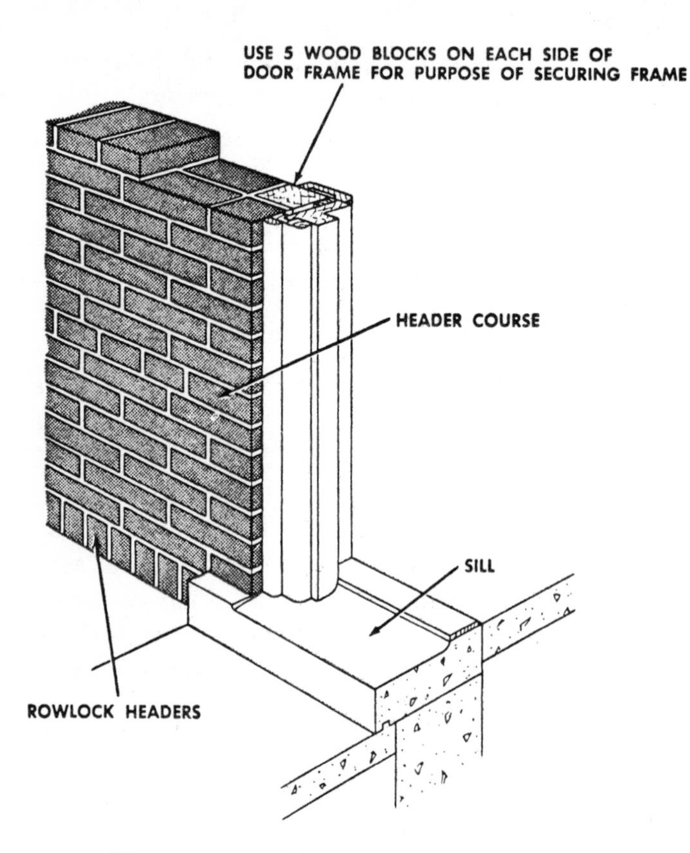

Figure 9-29. Construction at a door opening.

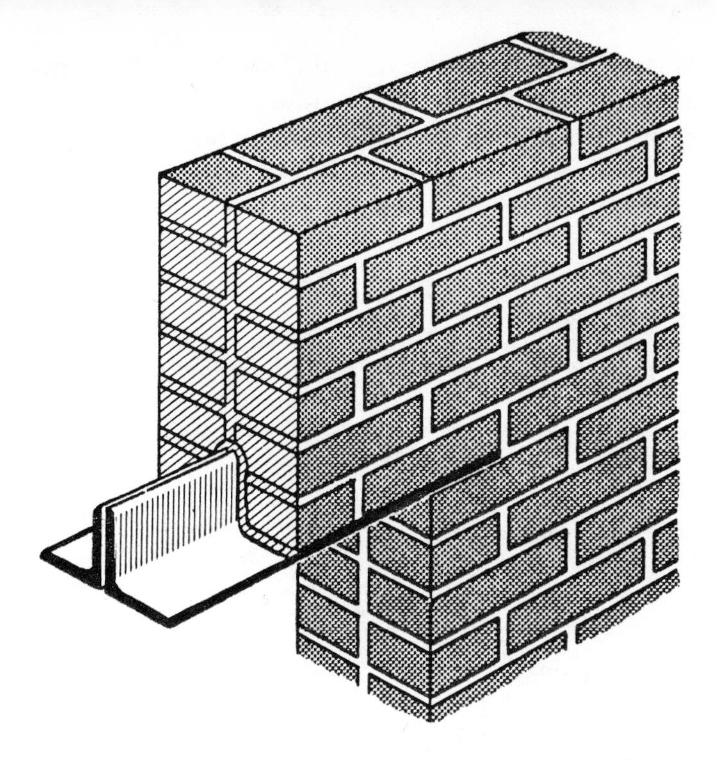

Figure 9–30. Lintels for an 8-inch wall.

by the proper installation of steel reinforcing bars. This is discussed in paragraph 9–34. Figures 9–30 and 9–31 illustrate some of the methods of placing lintels for different wall thicknesses. The relative placement and position is determined both by wall thickness and the type of window being used.

Usually the size and type of lintels required are given on drawings for the structure. When not given, the size of double-angle lintels required for various width openings in an 8-inch and 12-inch wall can be selected from table 9–5. Wood lintels for various width openings are also given in table 9–5.

Installation of a lintel for an 8-inch wall is shown in figure 9–30. The thickness of the angle for a two-angle lintel should be ¼ inch. This makes it possible for the two-angle legs that project up into the brick to fit exactly in the ½-inch joint between the face and backing-up ties of an 8-inch wall.

Corbeling

Corbeling consists of courses of brick set out beyond the face of the wall in order to form a

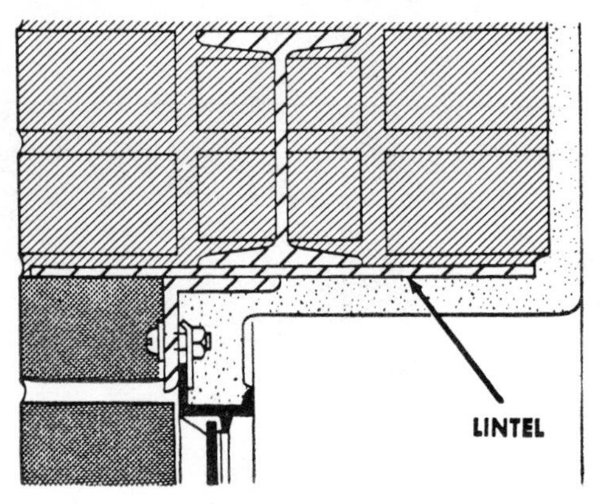

1. STEEL LINTEL

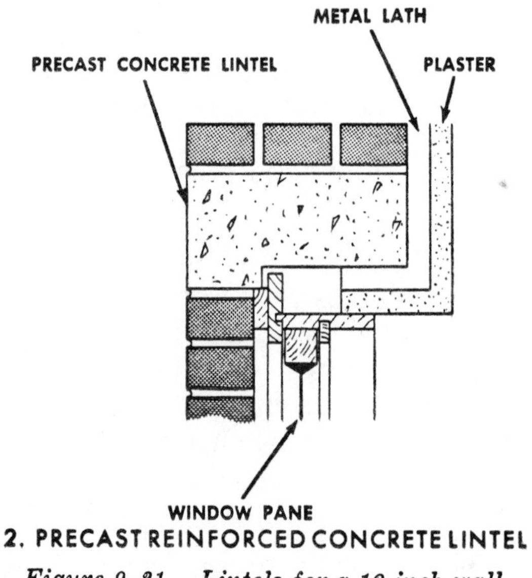

2. PRECAST REINFORCED CONCRETE LINTEL

Figure 9–31. Lintels for a 12-inch wall.

Table 9–5. Lintel Sizes

Wall thickness	Span						
	3 feet		4 feet* steel angles	5 feet* steel angles	6 feet* steel angles	7 feet* steel angles	8 feet* steel angles
	Steel angles	Wood					
8″	2–3 x 3 x ¼	2 x 8 2–2 x 4	2–3 x 3 x ¼	2–3 x 3 x ¼	2–3½ x 3½ x ¼	2–3½ x 3½ x ¼	2–3½ x 3½ x ¼
12″	2–3 x 3 x ¼	2 x 12 2–2 x 6	2–3 x 3 x ¼	2–3½ x 3½ x ¼	2–3½ x 3½ x ¼	2–4 x 4 x ¼	2–4 x 4 x 4¼

*Wood lintels should not be used for spans over 3 feet since they burn out in case of fire and allow the brick to fall.

self-supporting projection. This type of construction is shown in figure 9–32. The portion of a chimney that is exposed to the weather is frequently corbeled out and increased in thickness to improve its weathering resistance. Headers should also be used as much as possible in corbeling. It is usually necessary to use various-sized bats. The first projecting course may be a stretcher course if necessary. No course should extend out more than 2 inches beyond the course below it and the total projection of the corbeling should not be more than the thickness of the wall.

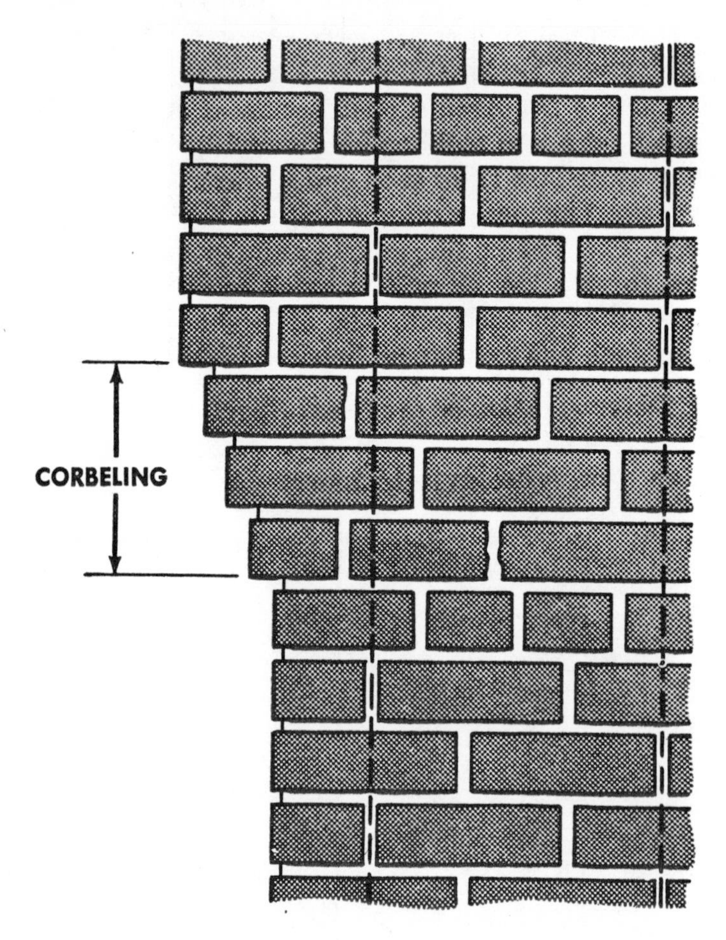

Figure 9–32. Corbeled brick wall.

Corbeling must be done carefully for the construction to have maximum strength. All mortar joints should be carefully made and completely filled with mortar. When the corbeled-out brick masonry is to withstand large loads, a qualified engineer should be consulted.

Brick Arches

Characteristics. If properly constructed, a brick arch can support a heavy load. The ability to support loads is derived primarily from it's curved shape. Several arch shapes can be used; the circular and elliptical shapes are most common (fig. 9–33). The width of the mortar joint is less at the bottom of the brick than it is at the top, and it should not be thinner than ¼ inch at any point. Arches made of brick must be constructed with full mortar joint's. As laying progresses, care must be taken to see that the arch does not bulge out of position.

Use of Temporary Support.

A brick arch is constructed on a temporary support that is left in position until the mortar has set. The temporary support is made of wood as shown in figure 9–34. The dimensions required are obtained from drawings. For arches up to 6 feet in span, ¾-inch plywood should be used for temporary supports. Two pieces cut to the proper curved shape are made and nailed to 2 by 4's placed between them. This will provide a wide-enough surface to support the brick adequately.

The temporary support should be held in position with wedges that can be driven out when the mortar has hardened enough for the arch to be self-supporting.

Laying the Arch. Construction of an arch is begun at the two ends or abutments of the arch.

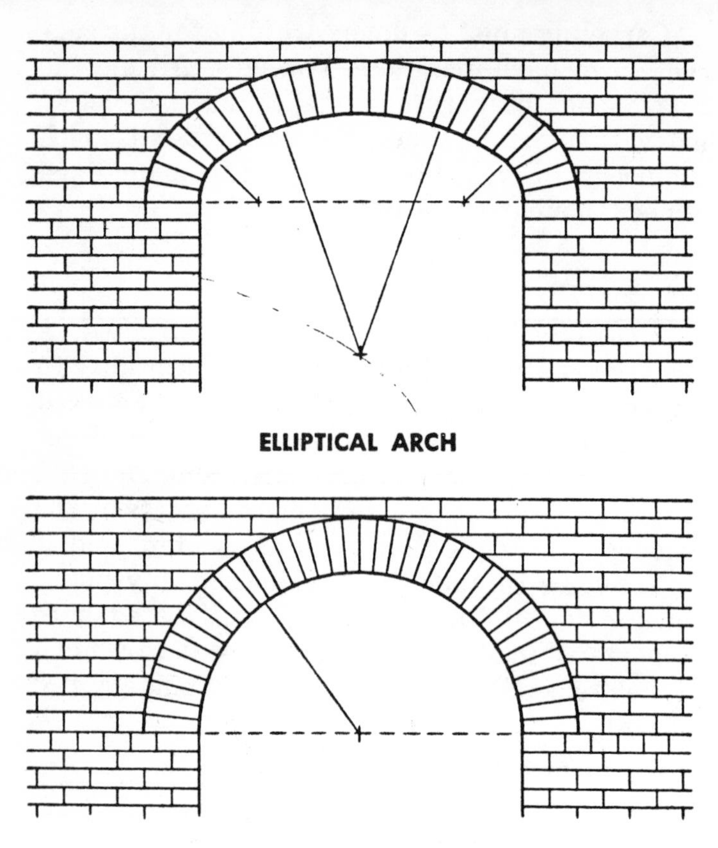

Figure 9–33. Types of arches.

The brick is laid from each end toward the center or crown. The key or middle brick is the last to be placed. There should be an odd number of brick in order for the key or middle brick to come at the exact center of the arch. The arch should be laid out in such a way that no brick need be cut.

Determination of the Brick Spacing. The best way to determine the number of brick required for an arch is to lay a temporary support on its side on level ground and set brick around it. Adjust the spacing until the key brick comes at the exact center of the arch. When this has been done,

Figure 9–34. Use of a templet in arch construction.

the position of the brick can be marked on the temporary support to be used as a guide when the arch is actually built.

Watertight Walls

The water that passes through brick walls does not usually enter through the mortar or brick but through cracks between brick and mortar. Sometimes these cracks are formed because the bond between the brick and mortar is poor. They are more apt to occur in head joints than in bed joints. To prevent this, some brick must be wetted. If the position of the brick is changed after the mortar has begun to set, the bond between the brick and mortar will be destroyed and a crack will result. Shrinkage of the mortar is also frequently responsible for the formation of cracks.

Both the size and number of cracks between the mortar and the brick can be reduced if the exterior face of all the mortar joints is tooled to a concave finish. All head joints and bed joints must be completely filled with mortar if watertightness is to be obtained.

Parging. A procedure found effective in producing a leakproof wall is shown in figure 9–35. The back of the brick in the face tier is plastered with not less than 3/8 inch of rich cement mortar before the backing bricks are laid. This is called parging or back plastering. Since parging should not be done over mortar protruding from the joints, all joints on the back of the face tier of bricks must be cut flush.

Membrane Waterproofing. Membrane waterproofing should be used if the wall is subject to considerable water pressure. The membrane, if properly installed, is able to adjust to any shrinkage or settlement without cracking. If the wall is to be subjected to considerable ground water or the surrounding soil is impervious, tile drains, or French drains if drainage tile is not available, should be constructed around the base of the wall (fig. 9–36).

Waterproof Coatings.

Bituminous mastic. For a foundation wall below ground level, two coats of bituminous mastic applied to the outside surface of the brick will yield satisfactory results. Asphalt or coal-tar pitch may be used and applied with mops.

Figure 9–35. Parging.

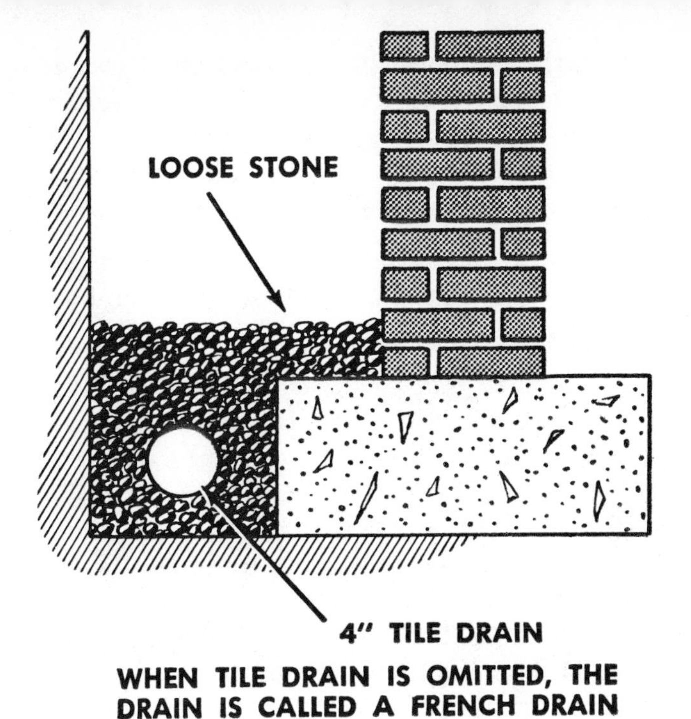

Figure 9–36. Drain around foundation.

Waterproof paints. The watertightness of brick walls above ground level is improved by the application of transparent waterproof paints such as a water solution of sodium silicate. Varnish is also effective. When used, these paints should be applied as specified by the manufacturer. Certain white and colored waterproofing paints are also available. In addition, good results have been obtained by the use of high-quality oil base paints.

Portland-cement paint. Portland-cement paint generally gives excellent results. The brick wall should be at least 30 days old before the portland-cement paint is applied and all efflorescence must be removed from the surface to be painted (para. 9–27). Type 2, class A portland-cement paint should be used (Fed. Spec. TT-P-21 (2)). Manufacturer's instructions for mixing and applying the paint are to be followed. Surfaces must be damp when the paint is applied. A water spray is the best means of wetting the surface. Whitewash or calcimine type brushes are used to apply the paint. Portland-cement paint can be applied with a spray gun but its rain resistance will be reduced.

Repairing Cracks. Before transparent waterproof paints or portland-cement paint is applied, all cracks should be repaired.

Cracks in mortar joints. Cracks in mortar joints are repaired by chipping out the mortar in the joint the full width of the joint to a depth of about 2 inches. The hole formed is carefully cleaned out by scrubbing with clean water. While the surface is still wet, a coating of cement mortar, made with sufficient water to form a thick liquid, should be applied. Before this coating sets, the hole should be filled with the prehydrated mortar described in paragraph 9–26 which is recommended for use in tuckpointing.

Cracks in brick. Cracks in the brick should be repaired the same as cracks in concrete (para 5–35, 5–36).

Fire-Resistant Brick

Purpose. Fire brick are manufactured for such uses as lining furnaces and incinerators. Their purpose is to protect the supporting structure or outer shell from intense heat. This outer shell may consist of common brick or, in some cases, steel, neither of which has good heat resistance.

Types of Fire-Resistant Brick. There are two types of fire-resistant brick:

Fire brick. These bricks are made from a special clay known as fire clay. They will withstand high temperatures and are heavier and usually larger than common brick. The standard size is 9 by 4½ by 2½ inches.

Silica brick. Silica brick should be used if resistance to acid gases is required. Silica brick should not be used if it is to be alternately heated and cooled. Most incinerators, therefore, should be lined with fire brick rather than silica brick.

Laying Fire Brick. Thin joints are of the utmost importance in laying fire brick. This is especially true when the brick are exposed to high temperatures such as those occurring in incinerators. The brick should be kept in a dry place until the time they are used.

Mortar. The mortar to be used in laying fire brick consists of fire clay mixed with water. The consistency of the mortar should be that of thick cream. Fire clay can be obtained by grinding used fire brick.

Laying procedure. The brick is dipped in the mortar in such a way that all faces except the top face are covered. The brick is then tapped firmly in place with a bricklayer's hammer. The joint between the brick should be as thin as possible and the brick should fit tightly together. Any cracks between the fire brick will allow heat to penetrate to the outside shell of the incinerator or furnace and damage it. The fire brick in one course lap those in the course below by one-half brick. The heat joints are thus staggered in the same way as they are staggered in the usual type of brick construction.

Laying Silica Brick. Silica brick are laid without mortar. They fit so closely that they fuse together at the joints when subjected to high temperatures. The head joints for silica brick are staggered, as for fire brick.

Special Types of Walls

Solid Walls. Many different types of walls may be built of brick. The solid 8- and 12-inch walls in common bond are the ones usually used for solid wall construction in the United States. The most important of the hollow walls are the cavity wall and the rowlock type wall.

Cavity Walls.

Cavity walls provide a means of obtaining a watertight wall that may be plastered without the use of furring or lathing. From the outside they appear the same as solid walls without header courses (fig. 9–37). No headers are re-

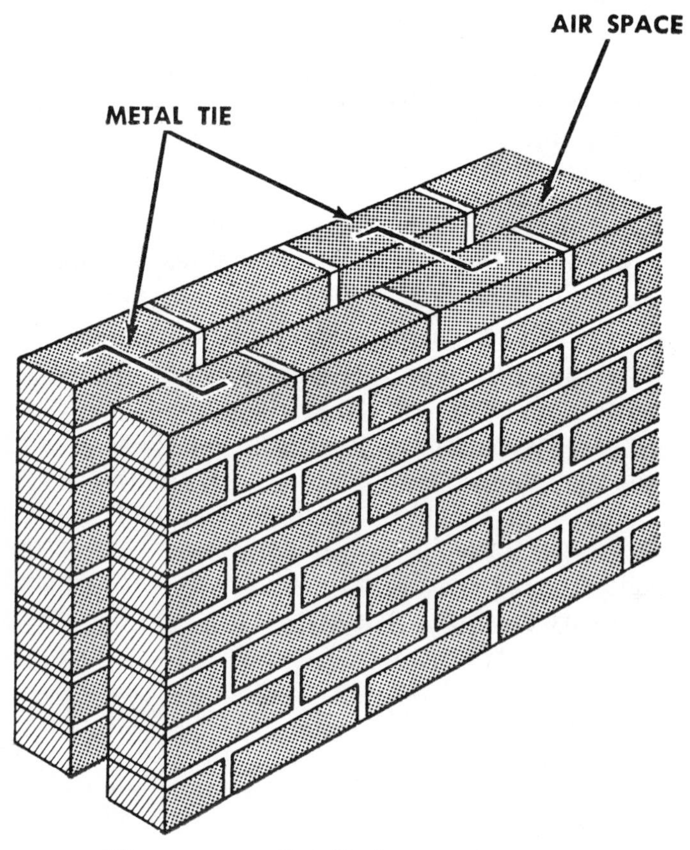

Figure 9–37. Details for a cavity wall.

quired because the two tiers of brick are held together by means of metal ties installed every sixth course and on 24-inch centers. To prevent waterflow to the inside tier, ties must be angled in a downward direction from the inside tier to the outside tier.

The 2-inch cavity between the two tiers of brick provides a space down which water that penetrates the outside tier may flow without passing through to the inside of the wall. The bottom of the cavity is above ground level and is drained by weep holes placed in the vertical joints between two bricks in the first course of the outer tier. These holes may be formed by leaving the mortar out of some of the vertical joints in the first course. The holes should be spaced at about 24-inch intervals. The air space also gives the wall better heat and sound insulation properties.

Rowlock-Back Wall.

One type of rowlock wall is shown in figure 9–38. The face tier of this wall has the same appearance as a common bond wall with a full header course every seventh course. The backing tier is laid with the brick on edge. The face tier and backing tier are tied together by a header course as shown. A 2-inch space is provided between the two tiers of brick, as for a cavity wall.

An all-rowlock wall is constructed with brick in the face and backing tier both laid on edge. The header course would be installed at every fourth course: three rowlock courses to every header course.

A rowlock wall is not as watertight as the cavity wall. Water is able to follow any crack present in the header course and pass through the wall to the inside surface.

Partition Walls. Partition walls that carry very little load can be made using one tier of brick only. This produces a wall 4 inches thick. A wall of this thickness is laid up without headers.

Laying Brick for Special Type Walls. Brick are laid in cavity walls and partition walls according to the procedure given in paragraph 9–8 for making bed joints, head joints, cross joints, and closures. The line is used the same as for a common bond wall. Corner leads for these walls

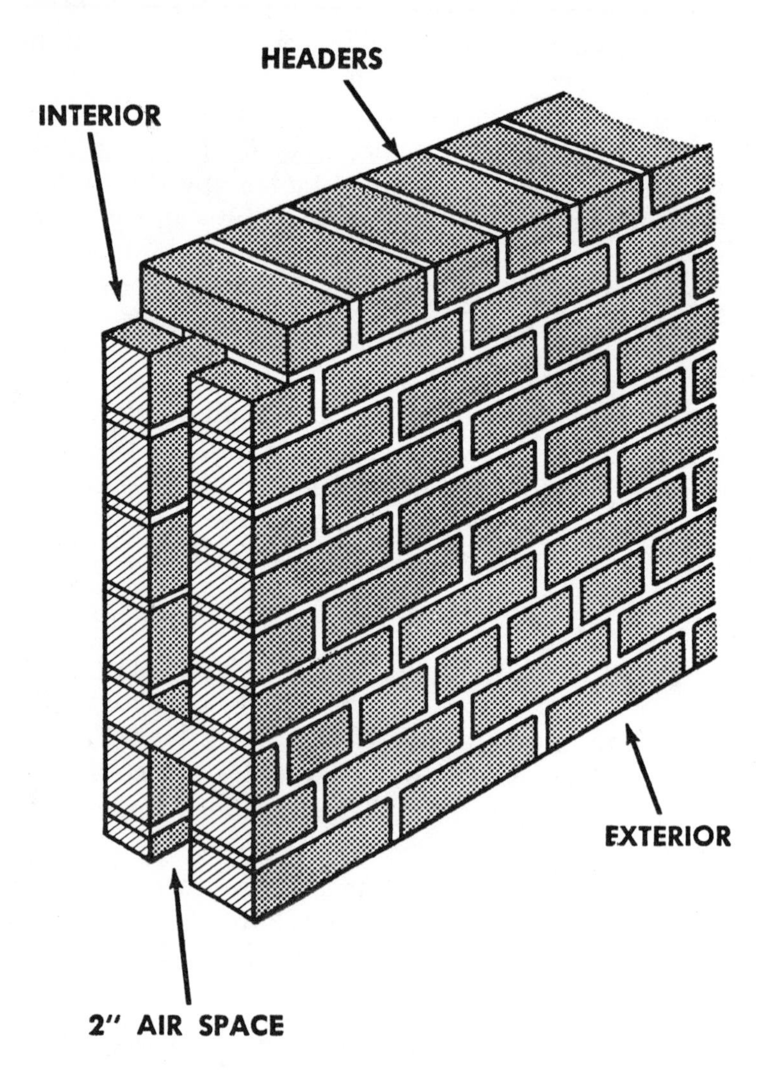

Figure 9–38. Details of a rowlock backwall.

are erected first and the wall between is built up afterward.

Manholes

Purpose.

Manholes are required for sewers so they can be cleaned and inspected. The size of the manhole required depends largely upon the size of the sewer. Manholes should be circular or oval, since this reduces the stress arising from both water and soil pressure. For small sewers on a straight line, a 4-foot diameter manhole is satisfactory. Details of a typical manhole are given in figure 9–39.

Both bottom and walls may be made of brick but normally the bottom is made of concrete because it is easier to cast the required shape in concrete rather than form it of brick. The walls of the manhole can be economically constructed if brick is used, eliminating the need for form work.

Required Wall Thickness. The thickness of walls required depends on the depth and the diameter of the manhole. An 8-inch thick wall should be used for manholes up to 8 feet in diameter and less than 15 feet deep. Manholes over 15 feet should be designed by a qualified engineer.

Manhole Construction.

Only headers are used for an 8-inch wall. A line cannot be used in the construction of the manhole wall. The level is employed to make sure that all brick in a course are at the same level. The straightedge or a straight, surfaced 2 by 4 may be used to span the manhole and the plumb rule placed on the 2 by 4 to determine whether the brick rises at the same level all around the manhole. Since the manhole wall will not be seen, some

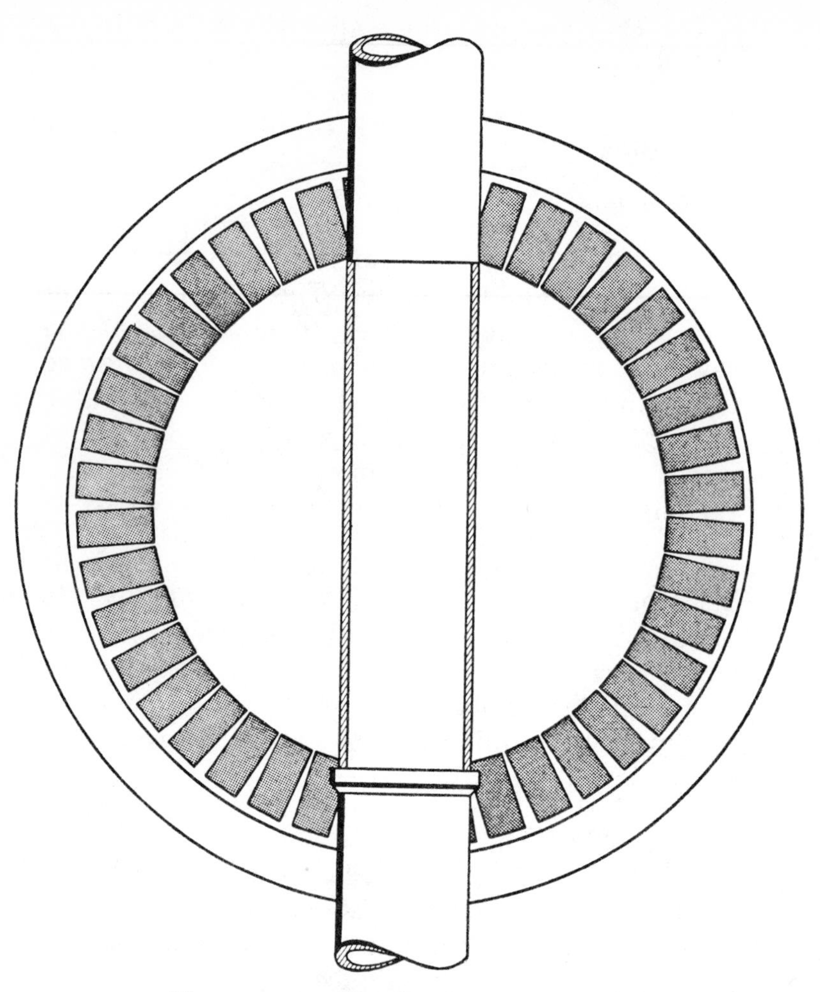

Figure 9–39. Details of a sewer manhole.

irregularities in brick position and mortar joint thickness are permissible. All joints in this type of wall are bed joints, or closure joints, made as described in paragraph 9–8.

Before the first course of brick is laid, a bed of mortar 1-inch thick should be placed on the foundation and the first course laid on this mortar bed.

To reduce the size of the manhole to that required for the manhole frame and cover, the

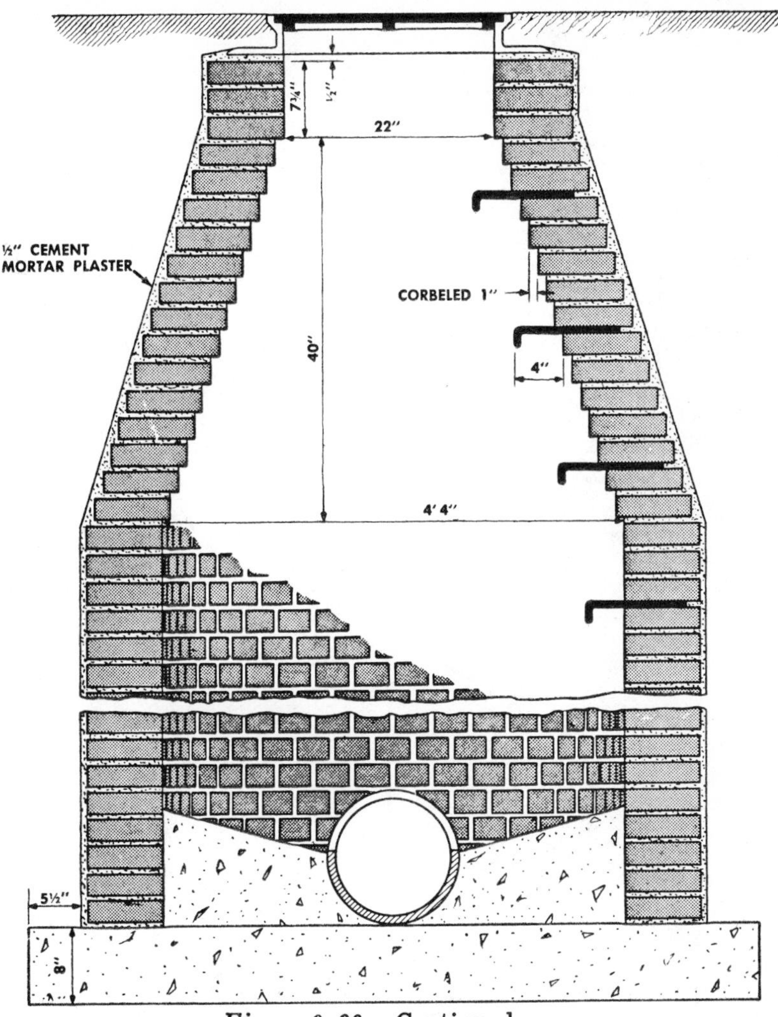

Figure 9–39.—Continued.

brick is corbeled inward as shown (2, fig. 9–39). One brick should not project more than 2 inches beyond the brick below it. Upon completion of the manhole wall, it is plastered on the outside with the same mortar used in laying the brick. The thickness of the mortar coating should be at least 3/8 inch.

The base of the manhole frame is placed in a 1-inch thick mortar bed spread on top of the

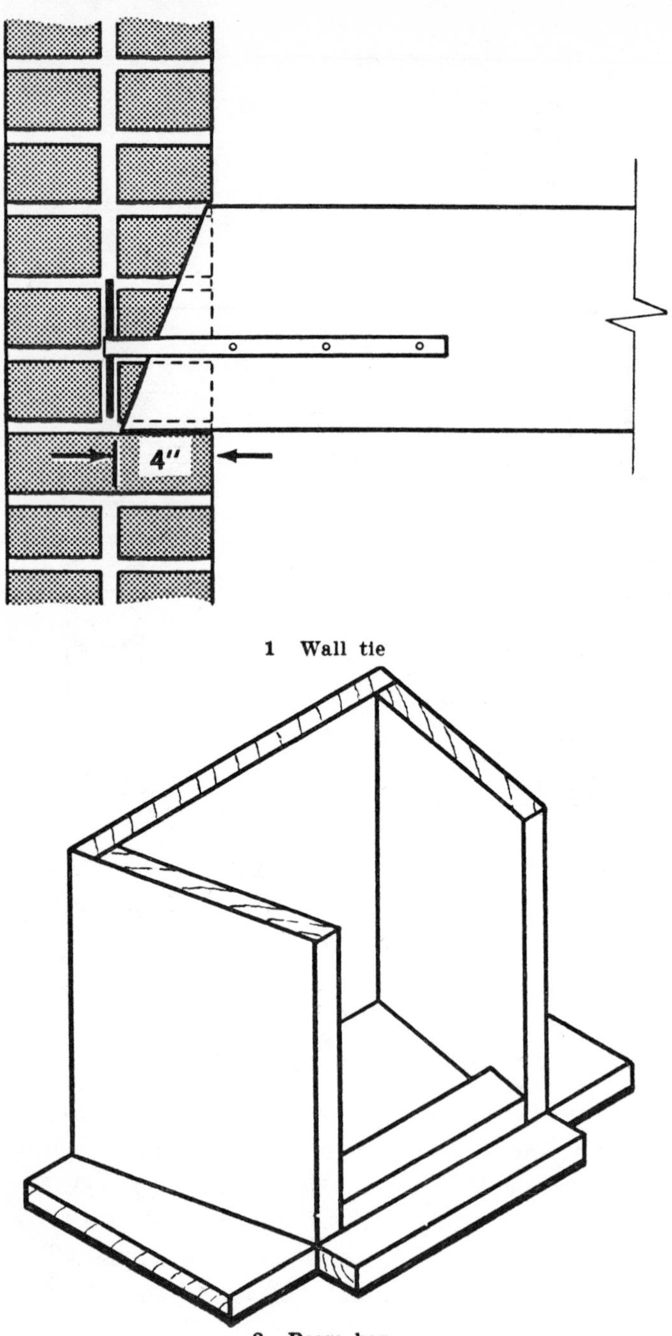

1 Wall tie

2 Beam box

Figure 9–40. Details of a wood beam supported by a brick wall.

manhole wall. The wrought-iron steps shown should be spaced at about 15 inches vertically and the embedded part placed in a cross mortar joint.

Method of Supporting Beams

When it is necessary to support a wood beam on a brick wall, it may be done as shown in 1, figure 9–40. Note the wall tie. If possible, the mortar should be kept away from the beam for dry rot may result if the wood is completely encased in mortar. This may be conveniently done by means of the wall box shown in 2, figure 9–40. The end of the beam should be cut at an angle which, in case of fire, will permit it to fall without damaging the wall above the beam. The beam should bear on the full width of the inside tier of brick for either an 8-inch or 12-inch wall.

If a steel beam is to be supported on a brick wall, a steel bearing plate set in mortar should be used under the beam. This bearing plate, when properly designed, reduces the tendency for the steel beam to crush the brick. The size of the bearing plate depends upon the size of beam and the load the beam carries. The method for determining the required thickness of bearing plates is beyond the scope of this manual.

Maintenance and Repair of Brick Walls

If a brick masonry wall is properly constructed, it requires little maintenance or repair. The proper repair of old masonry can be more expensive than the complete removal and replacement of the disintegrated portion. The use of good mortar, proper finishing of joints, and adequate flashing adds little to the initial cost and reduces the cost of maintenance.

Tuck-Pointing. Tuck-pointing consists of cutting out all loose and disintegrated mortar to a depth of at least ½ inch and replacing it with new. If leakage is to be stopped, all the mortar in the affected area should be cut out and new mortar placed according to instructions given in paragraph 9–21*e*. Tuck-pointing done as routine maintenance requires the removal of the defective mortar only.

Preparation of the Mortar Joint. All dust and loose material should be removed by brush or by means of a water jet after the cutting has been completed. A chisel with a cutting edge about ½-inch wide is suitable for cutting. If water is used in cleaning the joints, no further wetting is required. If not, the surface of the joint must be moistened.

Mortar for Tuck-Pointing. The mortar to be used for tuck-pointing should be portland-cement-lime, prehydrated type S mortar, or prehydrated prepared mortar made from type II masonary cement (para 7–3). The prehydration of mortar greatly reduces the amount of shrinkage. The procedure for prehydrating mortar is as follows: The dry ingredients for the mortar are mixed with just enough water to produce a damp mass of such consistency that it will retain its form when compressed into a ball with the hands. The mortar should then be allowed to stand for at least 1 hour and not more than 2 hours. After this has been done, it is mixed with the amount of water required to produce a stiff but workable consistency.

Filling the Joint. Sufficient time should be allowed for absorption of the moisture used in preparing the joint before the joint is filled with mortar. Filling the joint with mortar is called repointing and is done with a pointing trowel. The prehydrated mortar that has been prepared as above is packed tightly into the joint in thin layers about ¼ inch thick and finished to a smooth concave surface with a pointing tool. The mortar is pushed into the joint with a forward motion in one direction from a starting point to reduce the possibility of forming air pockets.

Cleaning New Brick Masonry

Cleaning New Brick Masonry. A skilled bricklayer is able to construct a masonry wall that is almost free of mortar stains. Most new brick walls, however, will need some cleaning.

Upon completion of the work, the large particles of mortar adhering to the brick are removed with a putty knife or chisel. Mortar stains are removed with acid prepared by mixing one part commercial muriatic acid with nine parts water. Pour the acid into the water. Before applying the acid, soak the surface thoroughly with water to prevent the mortar stain from being drawn into the pores of the brick.

The acid solution is applied with a long-handled, stiff-fiber brush. Proper precautions must be taken to prevent the acid from getting on hands, arms, and clothing. Goggles are worn to protect the eyes. An area of 15 to 20 square feet is scrubbed with acid and then immediately washed down with clear water. All acid must be removed before it can attack the mortar joint. Door and window frames must be protected from the acid.

Removing Efflorescence. Efflorescence is a white deposit of soluble salts frequently appearing on the surface of brick walls. These soluble salts are contained in the brick. Water penetrating the wall dissolves out the salts and when the water evaporates the salt remains. Efflorescence cannot occur unless both water and the salts are present. The proper selection of brick and a dry wall will keep efflorescence to a minimum. It may be removed, however, with the acid solution recommended for cleaning new walls. Acid should be used only after it has been determined that scrubbing with water and stiff brushes will not remove the efflorescence.

Cleaning Old Brick Masonry

Methods. Sand blasting, steam with water jets, and the use of cleaning compounds are the principal methods of cleaning old brick masonry. The process used depends upon the materials used in the wall and the nature of the stain. Many cleaning compounds that have no effect upon brick will damage mortar. Rough-textured brick is more difficult to clean than smooth-textured brick. Often it cannot be cleaned without removing part of the brick itself, and changing the appearance of the wall.

Sand Blasting. This method consists of blowing hard sand through a nozzle against the surface to be cleaned. Compressed air forces the sand through the nozzle. A layer of the surface is removed to the depth required to remove the stain. This is a disadvantage in that the surface is given a rough texture upon which soot and dust collect. Sand blasting usually cuts deeply into the mortar joints and it is often necessary to repoint them. After the sand blasting has been completed, it is advisable to apply a transparent waterproofing paint to the surface to help prevent soiling of the wall by soot and dust. Sand blasting is never done on glazed surfaces. A canvas screen placed around the scaffold used for sand blasting will make it possible to salvage most of the sand.

Cleaning by this method is accomplished by projecting a finely divided spray of steam and water at a high velocity against the surface to be cleaned. Grime is removed effectively without changing the texture of the surface.

The steam may be obtained from a portable truck-mounted boiler. The pressure should be from 140 to 150 pounds per square inch and about 12 boiler horse-power per cleaning nozzle is required. The velocity with which the steam and water spray hits the wall is more important than the volume of spray used.

A garden hose may be used to carry the water to the cleaning nozzle. Another garden hose supplies rinsing water. The operator experiments with the cleaning nozzle in order to determine the best angle and distance from the wall to hold the nozzle. The steam and water valves may also be regulated until the most effective spray is obtained. No more than a 3-foot-square area should be cleaned at one time. The cleaning should be done by passing the nozzle back and forth over the area, then rinsing it immediately with clean water before moving to the next space.

Sodium carbonate, sodium bicarbonate, or trisodium phosphate may be added to the cleaning water entering the nozzle to aid the cleaning action. The amount of salt remaining can be reduced considerably by washing the surface down with water before and after cleaning.

Hardened deposits that cannot be removed by steam cleaning should be removed with steel scrapers or wire brushes. Care must be taken not to cut into the surface. After the deposit has been removed, the surface should be washed down with water and steam cleaned.

Cleaning Compounds. There are a number of cleaning compounds that may be used, depending upon the stain to be removed. Most cleaning compounds contain material that will appear as efflorescence if allowed to penetrate the surface. This may be prevented if the surface to be cleaned is thoroughly wetted first. Whitewash, calcimine, and coldwater paints may be removed by the use of a solution of one part hydrochloric acid to five parts water. Fiber brushes are used to scrub the surface vigorously with the solution while the solution is still foaming. When the coating has been removed, the wall must be washed down with clean water until the acid is completely removed.

Paint Removers. Oil paint, enamels, varnishes, shellacs, and glue sizings should be removed with a paint remover applied with a brush and left on until the coating is soft enough to be scraped off with a putty knife. The following are effective paint removers:

Commercial paint removers. If these are used, follow the manufacturer's instructions.

Two pounds of trisodium phosphate in 1 gallon of hot water.

One and one-half pounds of caustic soda in 1 gallon of hot water.

Sand blasting or burning with a blowtorch.

Iron stains. The cleaning solution is prepared by mixing seven parts of lime-free glycerine with a solution of one part sodium citrate in six parts lukewarm water. Whiting or kieselguhr is added to make a thick paste to be applied to the stain with a trowel and scraped off when dry. The procedure is repeated until the stain in removed. After this the surface must be washed down with water.

Tobacco stains. The cleaning solution is prepared by dissolving 2 pounds of trisodium phosphate in 5 quarts of water. In an enameled pan, mix 12 ounces of chloride of lime in enough water to make a smooth thick paste. The trisodium phosphate solution is then mixed with the paste in a 2-gallon stoneware jar. When the lime has settled, the clear liquid is drawn off and diluted with an equal amount of water. A stiff paste is made by mixing the clear liquid with powdered talc; it is applied to the stain with a trowel and the surface washed.

Smoke stains. A smooth stiff paste made from trichlorethylene and powdered talc is effective in removing smoke stains. The solution container should be covered to prevent evaporation. If after several applications a slight stain still remains, the surface should be washed down and the procedure recommended for the removal of tobacco stains used. This solution should not be used in an unventilated space for the fumes are harmful.

Copper and bronze stains. These stains can be removed with a solution made by mixing one part of ammonium chloride (sal ammoniac) in dry form with four parts of powdered talc. Ammonia water is added and the solution stirred until a thick paste is obtained. This is applied to the stain with a trowel and allowed to dry. Several applications may be necessary after which the surface must be washed down with clear water.

Oil stains. Oil stains are effectively removed with a solution consisting of one part trisodium phosphate and 1 gallon of water to which enough whiting has been added to form a paste. The paste should be troweled over the stain in a layer ½ inch thick and allowed to dry for 24 hours. The paste is removed and the surface washed with clean water.

Flashing

Description.

Flashing is the impervious membrane placed at certain places in brick masonry for the purpose of excluding water or to collect any moisture that does penetrate the masonry and directing it to the outside of the wall. Flashing is installed at the head and sill of window openings and, in some buildings, at the intersection of the wall and roof. Where chimneys pass through the roof, the flashing should extend entirely through the chimney wall and turn up for a distance of 1 inch against the flue lining.

The edges of the flashing are turned up as shown in figure 9–41 to prevent drainage into the wall. Flashing is always installed in mortar joints. Drainage for the wall above the flashing is pro-

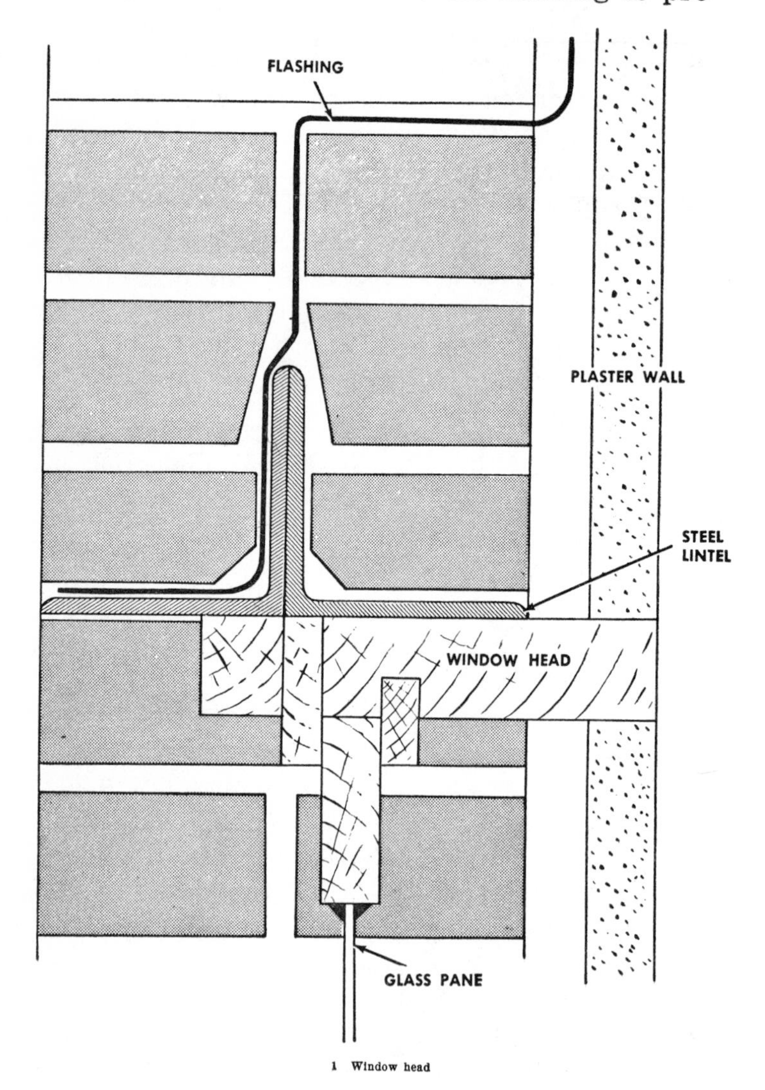

Figure 9–41. Flashing at window opening.

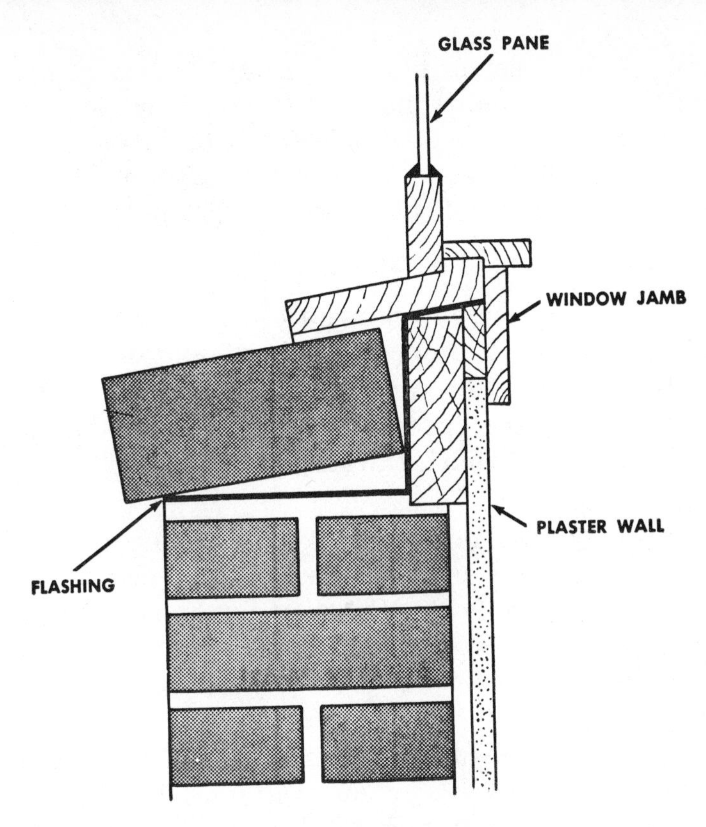

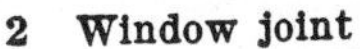

2 Window joint

Figure 9–41.—Continued.

vided by placing ¼-inch cotton-rope drainage wicks in the mortar joint just above flashing membrane at 18 inch spacings. Drainage may also be provided by holes left after dowels placed in the proper mortar joint are removed.

Flashing Materials. Copper, lead, aluminum, and bituminous roofing paper may be used for the flashing membrane. Copper is generally preferred but it will stain the masonry when it weathers. If this staining is undesirable, lead-coated copper should be used. Bituminous roofing papers are the cheapest but they are not as durable and may have to be replaced for permanent construction. The cost of replacement is many times the cost of installing high-quality flashing. Corrugated copper flashing sheets are available that produce a good bond with the mortar. These sheets have interlocking watertight joints at points of overlap.

Installation of Flashing.

In placing flashing, a ½-inch thick bed of mortar is spread on top of the brick and the flashing sheet pushed firmly down into the mortar. The brick or sill that goes on top of the flashing is forced into a ½-inch thick mortar bed spread on the flashing.

Details for the proper installation of flashing at the head and at the sill of a window are shown in figure 9–41. Note that at the steel lintel the flashing goes in under the face tier of brick, then back of the face tier, and finally over the top of the lintel.

The flashing required at the intersection of the roof and wall is shown in figure 9–42 and is always installed to prevent leakage between the roof and the wall. The upper end of the flashing is fitted and caulked into the groove of the raggle block.

Protection Against Freezing During Constructoin

Precaution. Leaky masonry walls can be attributed to the freezing of mortar before it set or to the lack of protection of materials and walls during cold weather construction. During cold weather, all materials and walls should be protected against freezing temperatures by:

Proper storage of materials.

Proper preparation of mortar.

Heating of masonry units.

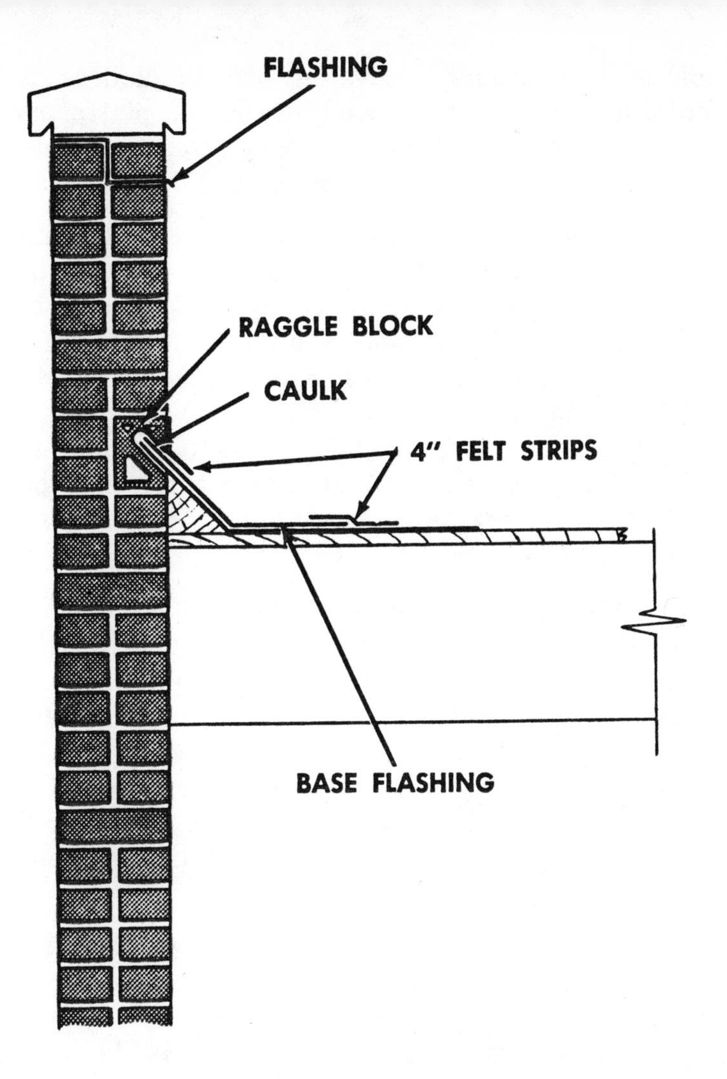

Figure 9–42. Flashing at intersection of roof and wall.

Laying precautions.
Protection of work.

Storing of Materials. Carelessness in storing materials during cold weather construction encourages poor workmanship, since the removal of ice and snow and the thawing of masonry units are absolutely necessary before construction proceeds. All masonry units and mortar materials should be thoroughly covered with tarpaulins or building paper and stored on plank platforms thick enough or raised high enough to prevent absorption of moisture from the ground.

Preparation of Mortar. Water and sand should be heated not to exceed 160° F. and the temperature of the mortar when used should be at least 70° F. and not more than 120° F. On small jobs where mortar boxes are used, they should be of steel and raised about a foot above the ground so that provisions can be made to supply some sort of heat to keep the mortar warm after mixing. *Salt should never be added to mortar to lower its freezing point.*

Heating of Masonry Units. To prevent sudden cooling of the warm mortar in contact with cold units it is recommended that all masonry units be heated to a unit temperature of about 40° F. when the outside temperature is below 18° F. The heating of masonry units requires careful planning. Where required, inside storage should be provided so that heat may be supplied at minimum expense.

Laying Precautions. In below freezing weather, brick having high rates of absorption should be sprinkled with *warm* water *just* before laying. Masonry units should never be laid on snow- or ice-covered beds because there will be little or no bond between the mortar and masonry units when the base thaws. Tops of unfinished walls should be kept carefully covered when work is not proceeding. If the covering is displaced and ice or snow collects on the wall top, it should be removed with live steam before the work continues.

Protection of Work. Protection of the masonry from freezing is necessary and will vary

*Table 9–6. Quantities of Material Required for Brick Walls**

Wall area sq ft	Wall thickness in inches							
	4 inches		8 inches		12 inches		16 inches	
	Number of bricks	Cu ft mortar	Number of bricks	Cu ft mortar	Number of bricks	Cu ft mortar	Number of bricks	Cu ft mortar
1	6.17	.08	12.33	.2	18.49	.32	24.65	.44
10	61.7	.8	123.3	2	184.9	3.2	246.5	4.4
100	617	8	1,233	20	1,849	32	2,465	44
200	1,234	16	2,466	40	3,698	64	4,930	88
300	1,851	24	3,699	60	5,547	96	7,395	132
400	2,468	32	4,932	80	7,396	128	9,860	176
500	3,085	40	6,165	100	9,245	160	12,325	220
600	3,712	48	7,398	120	11,094	192	14,790	264
700	4,319	56	8,631	140	12,943	224	17,253	308
800	4,936	64	9,864	160	14,792	256	19,720	352
900	5,553	72	10,970	180	16,641	288	22,185	396
1,000	6,170	80	12,330	200	18,490	320	24,650	440

*Quantities are based on ½-inch-thick mortar joint. For ⅜-inch-thick joint use 80 percent of these quantities. For ⅝-inch-thick joint use 120 percent.

with weather conditions. Each job is an individual problem. Job layout, desired rate of construction, and the prevailing weather conditions will determine the amount of protection and the type of heat necessary to maintain temperatures within the wall above freezing unit the mortar has set properly.

Temperature Variation. When the outside air temperature is below 40 °F., the temperature of the masonry when laid should be above 40° F. An air temperature of above 40° F. should be maintained on both sides of the masonry for at least 48 hours, if type M or S mortar is used, and for at least 72 hours if type N mortar is used. These periods may be reduced to 24 and 48 hours respectively, if high-early-strength cement is used. It is significant to note that the use of high-early-strength cement in mortars does not appreciably alter their rate of set. However, it does increase their rate of gaining strength, thereby providing greater resistance to further injury from freezing.

Quantities of Material Required for Brick Masonry

See table 9–6 for the quantity of brick and mortar required for various wall thicknesses.

REINFORCED BRICK MASONRY

Uses. Because the strength of brick masonry in tension is low, as compared with its compressive strength, reinforcing steel is used when tensile stresses are to be resisted. In this respect, brick masonry and concrete construction are identical. The reinforcing steel is placed in the horizontal or vertical mortar joints. Reinforced brick masonry may be used for beams, columns, walls, and footings in the same manner as reinforced concrete is used. Structures built of reinforced brick masonry have successfully resisted the effect of earthquake shocks intense enough to damage unreinforced brick structures severely. The design of reinforced brick masonry structures is similar to the design of reinforced concrete structures and is done by qualified engineers.

Materials.

Brick used for reinforced brick masonry is the same as that used for ordinary brick masonry. It should, however, have a compressive strength of at least 2,500 pounds per square inch.

The reinforcing steel is the same as the steel used to reinforce concrete and it is stored and fabricated in the same way. Hard-grade steel should not be used except in emergencies because many sharp bends are required in this type of construction.

Type N mortar (para 7–3) is used because of its high strength.

Wire for tying reinforcing steel should be 18-gage soft annealed iron wire.

Construction Methods for Reinforced Brick Masonry

Laying Reinforced Brick Masonry. Bricklaying is the same as for normal brick masonary. Mortar joint thickness is ⅛ inch more than the diameter of the steel bar used for reinforcing. This will allow 1/16 inch of mortar between the surface of the brick and the bar. When large steel bars are used, the thickness of the mortar joint will exceed ½ inch.

Placing the Steel. All reinforcing steel must be firmly embedded in mortar.

Horizontal bars are laid in a bed of mortar and pushed down until in position. More mortar is spread on top of the rods and smoothed out until a bed joint of the proper thickness can be made. The next course of brick is then laid in this mortar bed according to the procedure outlined for laying brick without reinforcing steel.

Stirrups for most reinforced brick beams must be of the shape shown in figure 9–43 in order to place them in the mortar joints. The lower leg is placed under the horizontal bars and in contact with them. Note that this may require a thicker joint at this point.

Vertical bars are placed in the vertical mortar joints. They are held in position by wood templets in which holes have been drilled at the proper bar spacing or by wiring to a horizontal bar. The brick is laid up around the vertical bars.

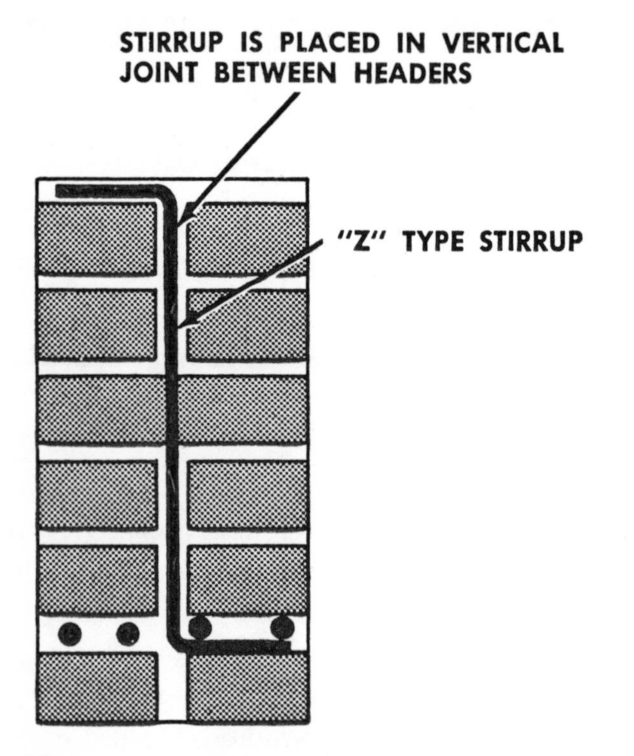

Figure 9–43. Reinforced brick masonry beam.

Horizontal and vertical bars need not be wired together as was recommended for reinforcing steel in concrete walls.

The minimum center-to-center spacing between parallel bars is 1½ times the bar diameter.

Forms.

Reinforced brick beams require form work for the same reason that reinforced concrete beams need form work. The form will consist only of a support for the bottom of the beam. No side form work is required. The form for the bottom is the same and is supported in the same way as recommended for concrete beams. No form work is required for walls, columns, or footings.

Where the beam joins a wall or another beam, the form should be cut ¼ inch short and the gap filled with mortar to allow for swelling of the lumber and to permit easy removal of the forms.

Form Removal. At least 10 days should elapse before the bottom form work for beams is removed.

Reinforced Brick Masonry Beams

Dimensions. The width and depth of beams depend upon brick dimensions, thickness of the mortar joints, and the load that the beam is required to support. Beam widths are usually the same as the wall thicknesses; that is, 4, 8, 12, and 16 inches. The depth should not exceed about three times the width.

Construction Procedures.

The first course of brick is laid on the form with full head joints but without a bed joint (fig. 9–43).

A bed of mortar about ⅛-inch thicker than the diameter of the horizontal reinforcing bars is spread on the first course of brick and the bars embedded in it as already described.

If stirrups are required, the leg of the stirrup is slipped under the horizontal bars as shown in figure 9–43. Care must be taken to get the stirrup in the center of the vertical mortar joint in which it is to be placed.

After the stirrups and the horizontal bars are in the proper position, spread additional mortar on the bed joint if necessary, and smooth the surface of the mortar. The mortar bed is now ready for the remaining courses which are laid in the usual way.

All of the brick in one course are laid before any brick in the next course are placed. This is necessary to insure a continuous bond be-

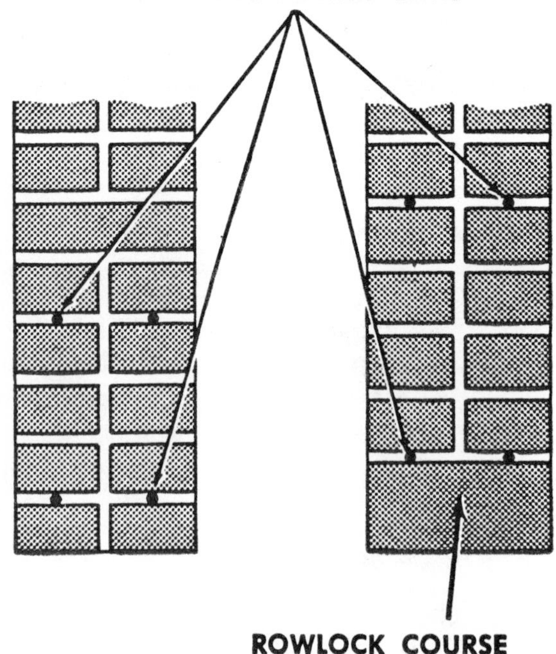

Figure 9–44. Reinforced brick masonry lintels.

tween the mortar and steel bars. It is frequently necessary to have three or four bricklayers working on one beam in order to get the bed joint mortar for the entire course spread, reinforcing steel placed, and brick laid before the mortar sets up.

Lintels. The proper placement of reinforcing steel in the brick wall above a window or door opening will serve the purpose of a lintel (para 9-18). The steel bars should be 3/8 inch in diameter or less if it is necessary to maintain a 1/2-inch thick mortar joint. The bars should extend 15 inches into the brick wall on each side of the opening and should be placed in the first mortar joint above the opening and also in the fourth joint above the opening (fig. 9–44). The lintel acts as a beam and needs a bottom form. The number and size bars required for different width wall openings are as follows:

Width of wall opening in feet	*Number and size of bars*
6	2 1/4-inch diameter bars
9	3 1/4-inch diameter bars
12	3 3/8-inch diameter bars.

Reinforced Brick Masonry Foundations

Large Footings. In large footings reinforcing steel is usually needed because of the tensil stresses that develop. As in all brick foundations, the first course of brick is laid in a bed of mortar about 1 inch thick that has been spread on the subgrade.

Wall Footings. A typical wall footing is shown in figure 9–45. The dowels extend above the footing; their purpose is to tie the footing and brick wall together. The No. 3 bars shown running parallel to the direction of the wall are used to prevent the formation of cracks perpendicular to the wall.

Column Footings. Column footings are usually square or rectangular and are reinforced as shown in figure 9–46. The dowels are needed to anchor the column to the footing and to transfer stress from the column to the footing. Note that both layers of horizontal steel are placed in the same mortar joint. This is not necessary and, if large bars are used, one layer of steel should be placed in the second mortar joint above the bottom. If this is done, the spacing between the bars in the upper layer of steel must be reduced.

Reinforced Brick Masonry Columns

Columns. The load-carrying capacity of brick columns is increased when they are reinforced with steel bars. There should be at least 1½ inches of mortar or brick covering the reinforcing bars and these bars should be held in place with ⅜-inch diameter steel hoops or ties as shown in figure 9–47. When possible, the hoops or ties should be circular rather than rectangular or square. The ends of the hoop or tie should be lap-welded together or bent around a reinforcing bar as shown

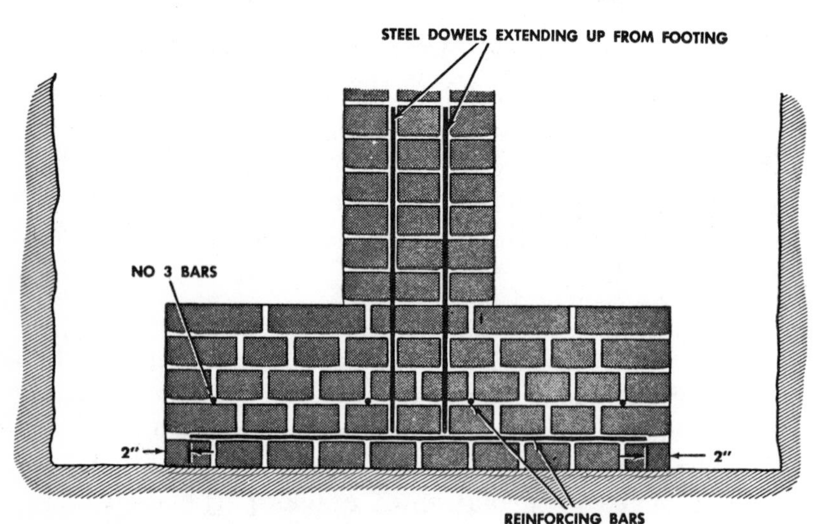

Figure 9–45. Reinforced masonry wall footing.

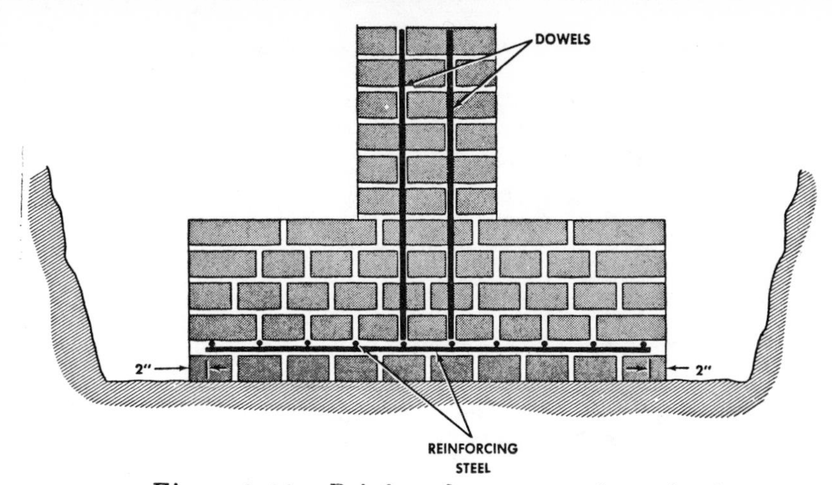

Figure 9–46. Reinforced masonry column footing.

in figure 9–47. Hoops should be installed at every course of brick.

Method of holding steel in place. After the footings are completed, the column reinforcing steel is tied to the dowels projecting from the footing. The required number of hoops is then slipped over the longitudinal reinforcing bars and temporarily fastened to these bars some distance above the level at which brick are being laid but within reach of the bricklayer. It is not necessary for the hoops to be held in position by wiring them to the longitudinal reinforcing. The tops of the longitudinal reinforcing bars are held in position by means of a wood templet or by securely tying them to a hoop placed near the top of the column.

Laying the brick. The brick are laid as described in paragraph 9–8. The hoops are placed in a full bed of mortar and the mortar smoothed out before the next course of brick is laid. Brick bats may be used in the core of the column or where it is inconvenient or impossible to use full-size brick. After all the brick in a course are laid, the core and all remaining space around the reinforcing bars is filled with mortar. Any bats required are then pushed into the mortar until com-

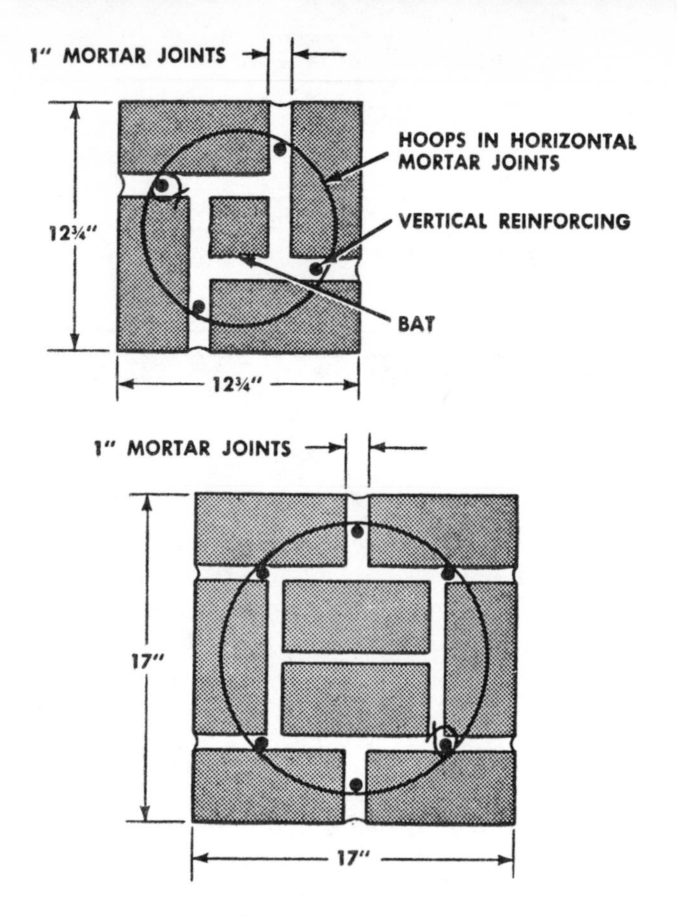

Figure 9–47. Reinforced brick masonry columns.

pletely embedded. The next mortar bed is now spread and the process repeated.

Walls. Reinforcing steel for walls consists of both horizontal and vertical bars and is placed as discussed in paragraph 9–33 .The vertical bars are wired to the dowels projecting up from the footing below and are placed in the mortar joint between tiers of brick.

Laying the brick. As the brick are laid, all space around the bars is filled with mortar. Otherwise, the wall is constructed as described in paragraphs 9–8 and 9–34.

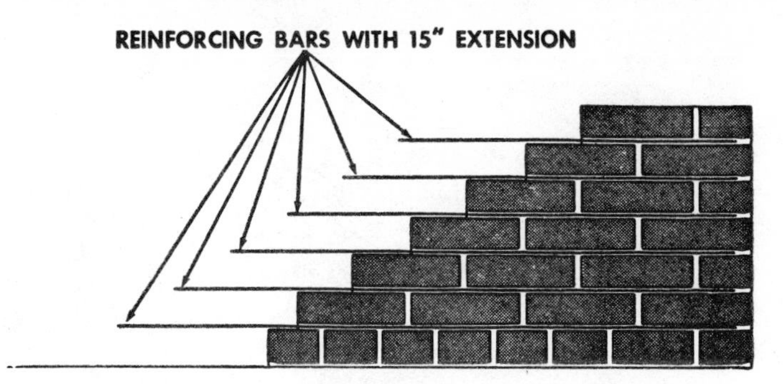

Figure 9–48. Corner lead for reinforced brick masonry wall.

Corner leads. In the construction of corner leads, bars should be placed in the corner as shown in figure 9–48. The extension is 15 inches and the bar size should be the same as that used for horizontal bars in the rest of the wall. The horizontal bars in the remainder of the wall lap these corner bars by the same 15 inches. As for beams, all the brick in one course between corner leads are laid before any other brick are laid. This is necessary since the entire reinforcing bar must be embedded in mortar at the same time.

STRUCTURAL CLAY TILE MASONRY

Types of Structural Clay Tile

Description.

Hollow units made of burned clay or shale are normally referred to as clay tile. Several common types of clay tile and their dimensions are shown in figure 9–49. These masonry units are made by forcing a plastic clay through special dies The tiles is then cut to size and burned in the same manner as brick. The amount of burning depends upon the grade of tile being manufactured.

The hollow spaces in the tile are called cells. The outside wall of the tile is called the shell

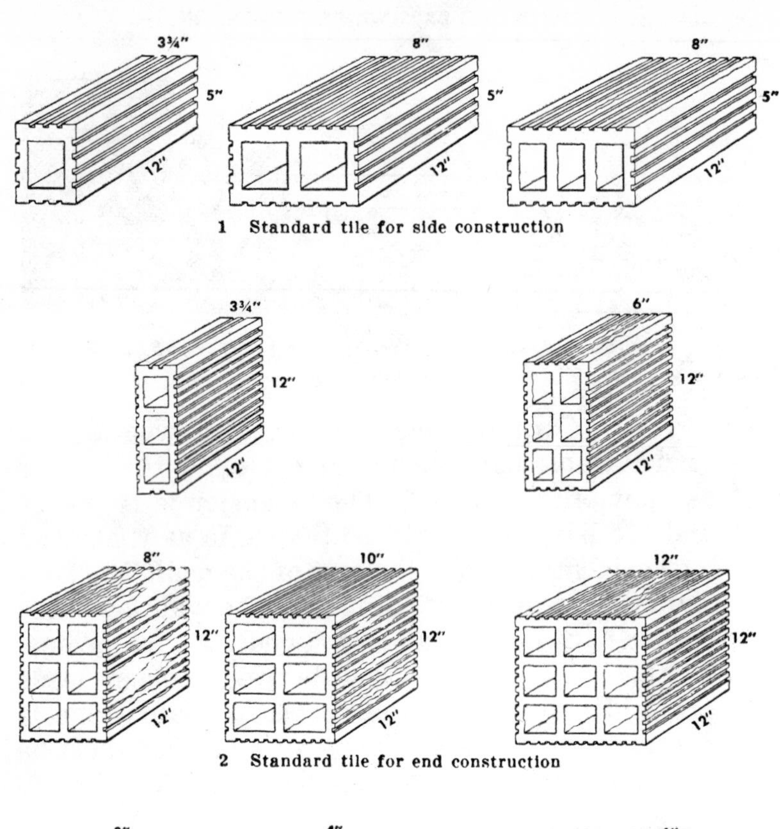

1 Standard tile for side construction

2 Standard tile for end construction

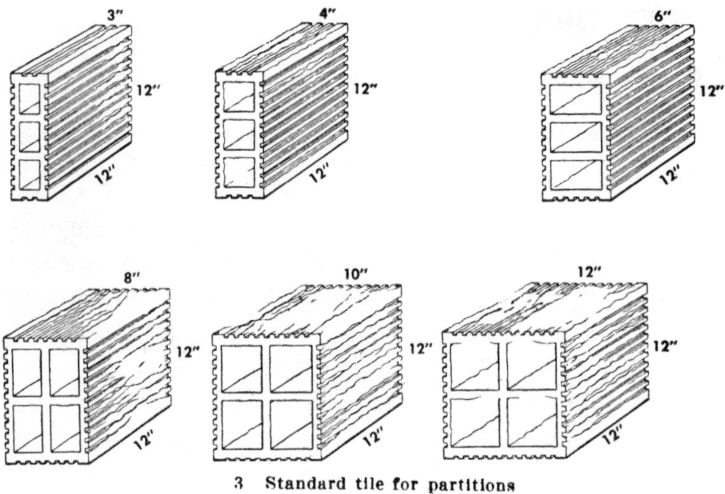

3 Standard tile for partitions

Figure 9–49. Types of structural clay tilė.

and the partitions that divide the title into cells are called webs. The shell should be at least ¾-inch thick and the web ½-inch thick.

In side-construction tile the cells are horizontal while in end-construction tile the cells are placed vertically. There is no particular reason to prefer one type to the other. Both side-construction and end-construction tile are made in the following structural clay tile classifications.

Structural Clay Load-Bearing Wall Tile. Tile in this classification may be divided into two types depending upon use.

Tile used for the construction of exposed or faced load-bearing walls. This tile is designed to carry the entire load including the facing material. The facing may consist of stucco, plaster, or other suitable material.

Tile used for backing up load-bearing and non-loadbearing brick walls. The facing or outer tier of brick in these walls is bonded to the backing tile by headers. The wall load is supported by both the facing and the backing. The inside face of this tile is scored, making it possible to plaster it without using lath.

ASTM Classification. The American Society for Testing and Materials gives two grades of structural load-bearing tile based on resistance to weathering.

Grade LB. This grade is suitable for general use on masonry construction where not exposed to weathering or for use in masonry exposed to weathering provided it is protected with at least 3 inches of facing.

Grade LBX. This tile may be used in masonry exposed to weathering with no facing material required.

Structural Clay Non-Load-Bearing Tile. This classification includes:

Partition tile used in the construction of non-load-bearing partitions or for backing non-load-bearing brick walls.

Tile used for lining the inside of a wall in order to provide a surface that may be plastered. These tiles also provide an air space between the plaster and the wall.

Tile used to fireproof steel columns and beams.

Structural Clay Facing Tile. The ASTM gives two classes of structural clay facing tile based on shell thickness and called standard and special duty. The surface finish of this tile closely resembles the surface finish of face brick. There are two types in each class.

Type FTX. This tile is the better in appearance and is the easier to clean.

Type FTS. This title is inferior in quality to type FTX but is suitable for general use where some defects in surface finish are not objectionable.

*Table 9–7. Quantities of Material Required for Side Construction Hollow Clay Tile Walls**

Wall thickness—tile size—wall area, sq ft.	4 inches—4 x 5 x 12		8 inches—8 x 5 x 12	
	Number of tile	Cu ft. mortar	Number of tile	Cu ft. mortar
1	2.1	.045	2.1	.09
10	21	.45	21	.9
100	210	4.5	210	9.0
200	420	9.0	420	18
300	630	13.5	630	27
400	840	18.0	840	36
500	1,050	22.5	1,050	45
600	1,260	27.0	1,260	54
700	1,470	31.5	1,470	63
800	1,680	36.0	1,680	72
900	1,890	40.5	1,890	81
1,000	2,100	45.0	2,100	90

*Quantities are based on ½-inch thick mortar joint.

Structural Glazed Facing Tile. The exposed surface of glazed tile has either a ceramic glaze, a salt glaze, or a clay coating. These tiles are used where stainproof, easily cleaned surfaces are desired. They can be obtained in many colors and produce a durable wall with a pleasing appearance.

Special Units. In addition to the standard stretcher units shown in figure 9–49, special units are available for use at window and door openings and at corners. A manufacturer's catalog should be consulted if special units are necessary.

Quantities of Materials Required. The amount of materials required for structural clay tile walls is given in tables 9–7 and 9–8.

Physical Characteristics

Strength. The compressive strength of the individual tile depends upon the materials used and upon the method of manufacture in addition to the thickness of the shells and webs. A minimum compressive strength of tile masonry of 300 pounds per square inch based on the gross section may be expected. The tensil strength of structural clay tile masonry is small. In most cases, it is less than 10 percent of the compressive strength.

Abrasion Resistance. As for brick, the abrasion resistance of clay tile depends primarily upon its compressive strength. The stronger the tile, the greater its resistance to wearing. The abrasion resistance decreases as the amount of water aborbed increases.

Weather Resistance. Structural clay facing tile has excellent resistance to weathering. Freezing and thawing action produces almost no deterioration. Tile that will absorb no more than 16 percent of their weight of water have never given

*Table 9–8. Quantities of Material Required for End Construction Hollow Clay Tile Walls**

Wall thickness	4 inches		6 inches		8 inches		10 inches	
Tile size	4 x 12 x 12		6 x 12 x 12		8 x 12 x 12		10 x 12 x 12	
Wall area, sq. ft.	Number of tile	Cu. ft. mortar	Number of tile	Cu. ft. mortar	Number of tile	Cu. ft. mortar	Number of tile	Cu. ft. mortar
1	.93	.025	.93	.036	.93	.049	.93	.06
10	9.3	.25	9.3	.36	9.3	.49	9.3	.4
100	93	2.5	93	3.6	93	4.9	93	6
200	186	5.0	186	7.2	186	9.8	186	12
300	279	7.5	279	10.8	279	14.7	279	18
400	372	10.0	372	14.4	372	19.6	372	24
500	465	12.5	465	18.0	465	24.5	465	30
600	558	15.0	558	21.6	558	29.4	558	36
700	651	17.5	651	25.2	651	34.3	651	42
800	744	20.0	744	28.8	744	39.2	744	48
900	837	22.5	837	32.4	837	44.1	837	54
1,000	930	25.0	930	36.0	930	49.0	930	60

*Quantities are based on ½-inch-thick mortar joint.

unsatisfactory performance in resisting the effect of freezing and thawing action. Only portland-cement-lime mortar or mortar prepared from masonary cement should be used if the masonry is exposed to the weather.

Heat- and Sound-Insulating Properties. Walls containing structural clay tile have better heat-insulating qualities than do walls composed of solid units, due to dead air space that exists in tile walls. The resistance to sound penetration of this type of masonry compares favorably with the resistance of solid masonry walls but it is somewhat less.

Fire Resistance. The first resistance of tile walls is considerably less than the fire resitance of solid masonry walls. It can be improved by applying a coat of plaster to the surface of the wall. Partition walls of structural clay tile 6 inches thick will resist a fire for 1 hour provided the fire produces a temperature of 1,700° F. after burning for 1 hour.

Weight. The solid material in structural clay tile weighs about 125 pounds per cubic foot. Since the tile contains hollow cells of various sizes, the weight of tile varies, depending upon the manufacture and type. A 6-inch tile wall weighs approximately 30 pounds per square foot, while a 12-inch tile wall weighs approximately 45 pounds per square foot.

Uses for Structural Clay Tile

Exterior Walls. Structural clay tile may be used for exterior walls of either the load-bearing or non-load-bearing type. It is suitable for both below-grade and above-grade construction.

Partition Walls. Non-load-bearing partition walls of from 4- to 12-inch thickness are frequently made of structural clay tile. These walls are easily built, light in weight, and have good heat-and sound-insulating properties.

Backing for Brick Walls. Figure 9–50 illustrates the use of structural clay tile as a backing unit for a brick wall.

Mortar Joints for Structural Clay Tile

End-Construction Units. In general, the procedure for making mortar joints for structural clay tile is the same as for brick.

Bed Joint. The bed joint is made by spreading a 1-inch thickness of mortar on the shell of the bed tile but not on the webs, as shown in 1, figure 9–51. The mortar should be spread for a distance of about 3 feet ahead of the laying of the tile. The position of the tile above does not coincide with the position of the tile below since the head joints are to be staggered. The web of the tile above will not contact the web of the tile below and any mortar placed on these webs is useless.

Head joint. The head joint is formed by spreading plenty of mortar along each edge of the tile, as shown in 2, figure 9–51, and then pushing

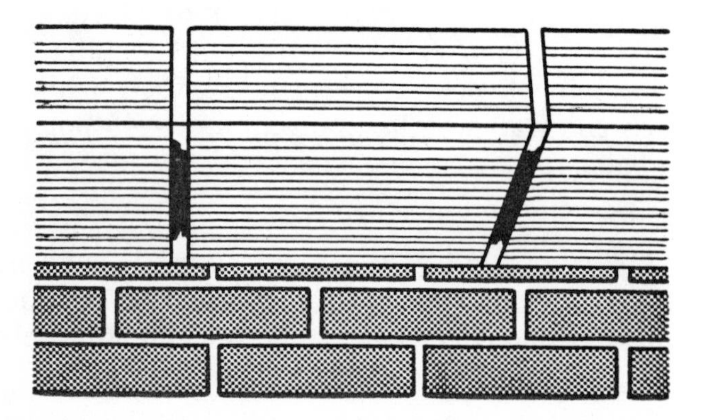

Figure 9–50. Structural clay tiles used as a backing unit.

the tile into the mortar bed until in its proper position. Enough mortar should be used to cause excess mortar to squeeze out of the joints. This excess mortar is cut off with a trowel. The head joint need not be a solid joint as recommended for head joints in brick masonry unless the joint is to be exposed to the weather. Clay tile units are heavy, making it necessary to use both hands when placing the tile in position in the wall. The mortar joint should be about ½-inch thick, depending upon the type of construction.

Closure joints. The procedure recommended for making closure joints in brick masonry should be used.

Side-Construction Units.

Bed joint. The bed joint is made by spreading the mortar to a thickness of about 1 inch for a distance of about 3 feet ahead of the laying of the tile. A furrow need not be made as is required for brick.

Head joint.

Method A. As much mortar as will adhere is spread on both edges of tile as shown in 1, figure 9–52. The tile is then pushed into the mortar bed against the tile already in place until in its proper position. Excess mortar is cut off.

Method B. As much mortar as will adhere is placed on the interior edge of the tile already in place and on the opposite edge of the unit being placed. This is shown in 2, figure 9–52. The tile is then shoved in place and the excess mortar cut off.

Mortar joint thickness. The mortar joints should be about ½ inch thick, depending upon the type of construction.

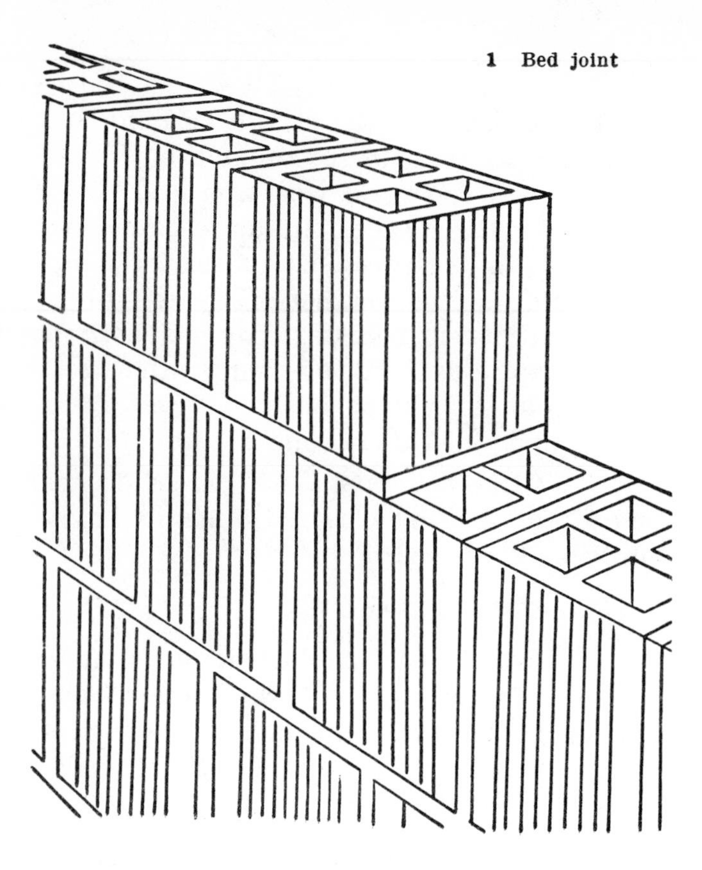

Figure 9–51. Laying end-construction tile.

Eight-Inch Wall With Four-Inch Structural Tile Backing

For this wall there will be six stretcher courses between the header courses. The backing tile is side-construction 4- by 5- by 12-inch tile. They are 4 inches wide, 5 inches high and 12 inches long. The 5-inch height is equal to the height of two brick courses and a ½-inch mortar joint. These tiles are laid with a bed joint such that the top of the tile will be level with every second course of brick. The thickness of the bed joint therefore depends upon the thickness of the bed joint used for the brick.

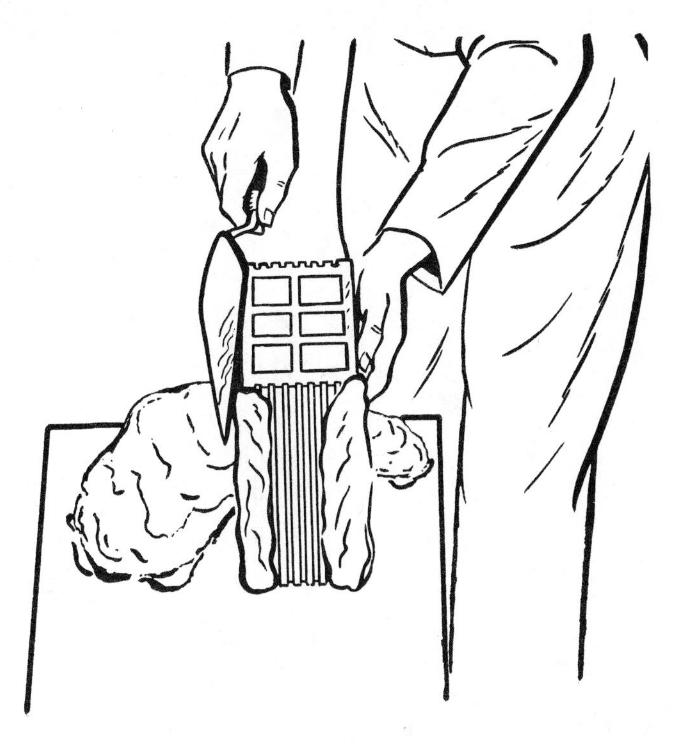

2 Head joint

Laying Out the Wall. The first course of the wall is temporarily laid out without mortar as recommended for solid brick walls. This will establish the number of brick required for one course.

Laying the Corner Leads. As shown in 1, figure 9–53. the first course of the corner lead is identical to the first course of the corner lead for a solid 8-inch brick wall except that one more brick is laid next to the brick *p*, 5, figure 9–22. All the brick required for the corner lead are laid before any tile is placed. The first course of tile is shown in 2, figure 9–53 and the completed corner lead is shown in 3, figure 9–53. The remainder of the wall is completed as described in paragraph 9–16.

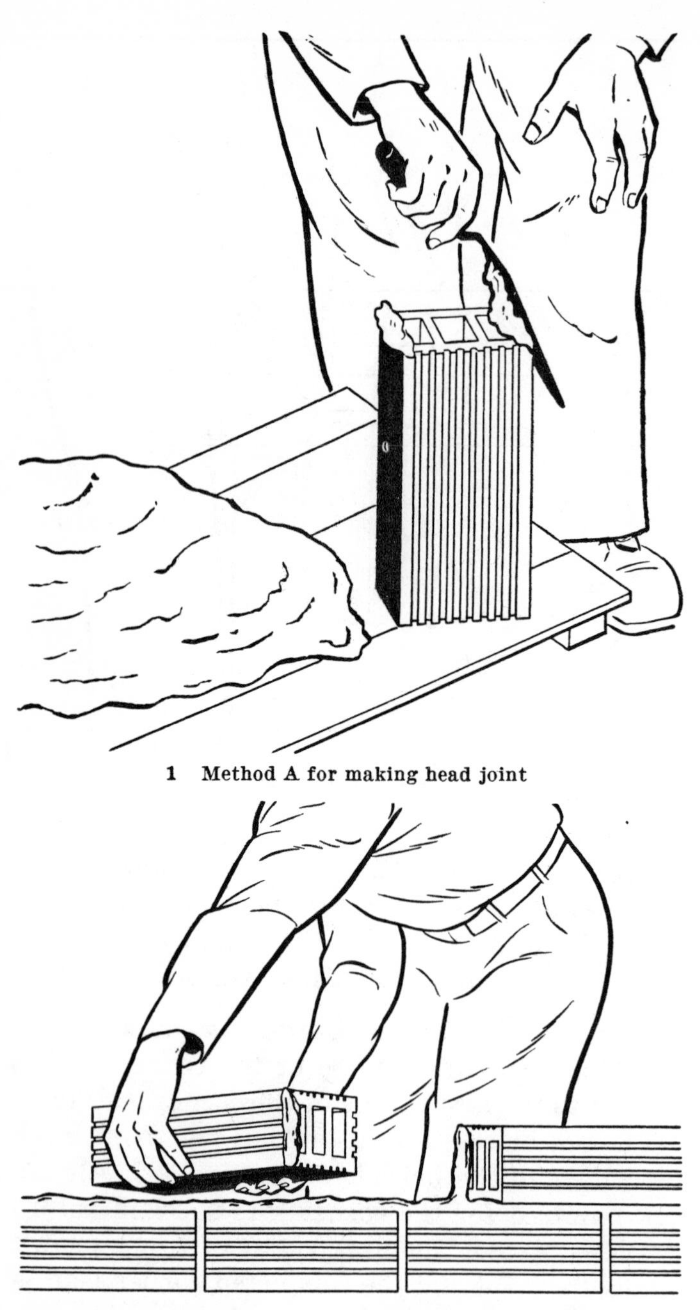

1 Method A for making head joint

2 Method B for making head joint

Figure 9–52. Laying side-construction tile.

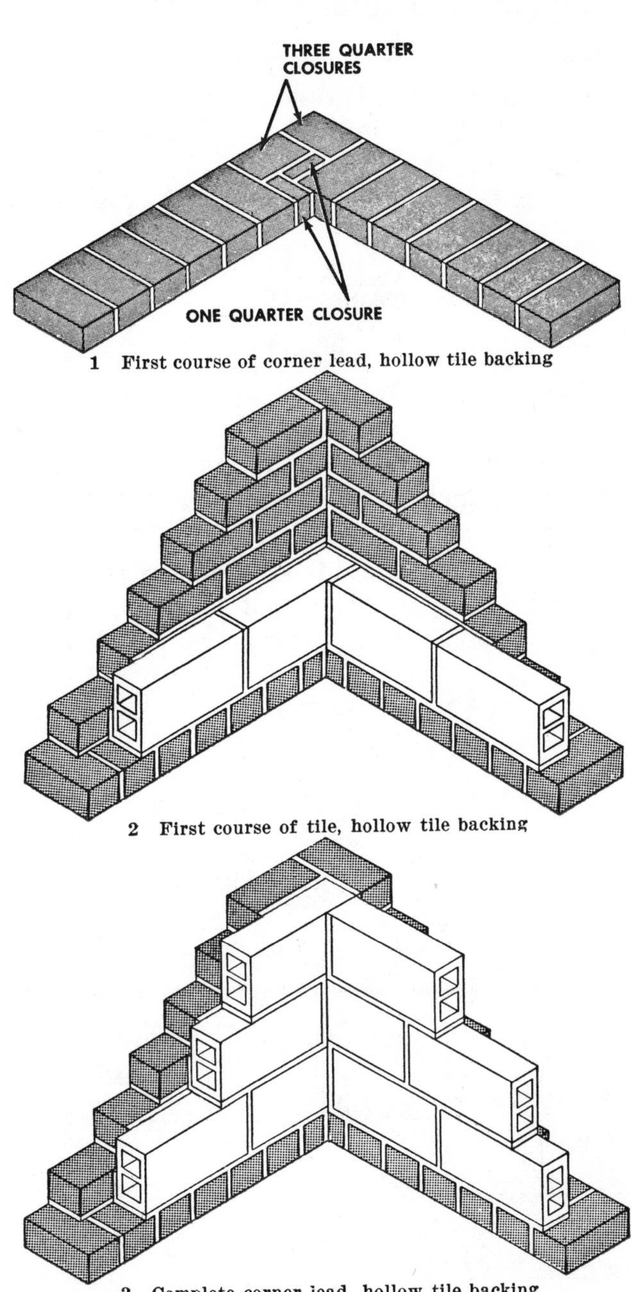

1 First course of corner lead, hollow tile backing

2 First course of tile, hollow tile backing

3 Complete corner lead, hollow tile backing

Figure 9–53. Corner lead hollow-tile backing.

Eight-Inch Structural Clay Tile Wall

This wall is constructed of 8- by 5- by 12-inch tile. The length of the tile is 12 inches, the width is 8 inches, and the height is 5 inches. A 2- by 5- by 8-inch soap is used at the corners as shown in figure 9–54. Half-lap bond is used as indicated.

Laying Out the Wall. This is done as described in paragraph 9–16*a*.

Laying the Corner Leads. Tiles a and b (fig. 9–54) are laid first, then c and d. The level is checked as they are laid. Tiles e and f are laid and their level checked. Tile b must be laid so that it projects 6 inches from the inside corner as shown to provide for the half-lap bond. Corner tiles such as b, g, and g, should be end-construction tile in order to avoid exposure of the open cells at the face of the wall, or a thin end-construction tile,

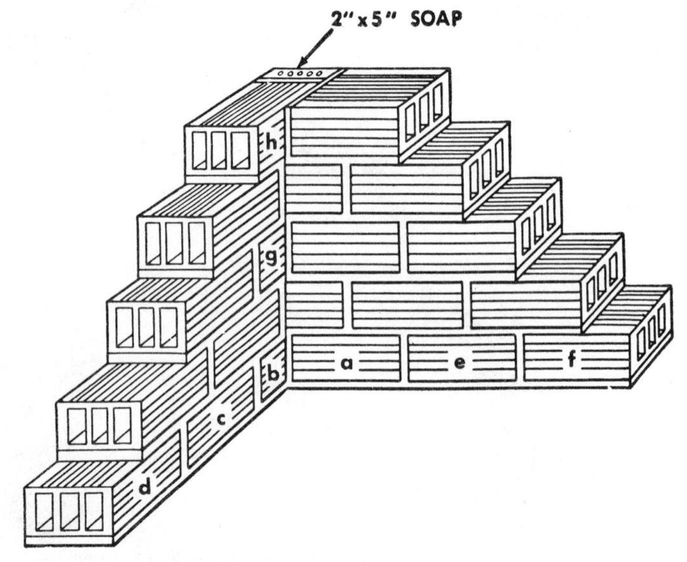

Figure 9–54. Eight-inch structural clay tile wall.

known as a “soap”, may be used at the corner as shown in figure 9–54. The remainder of the tile in the corner is then laid, and the level of each is checked. After the corner leads are erected, the wall between is laid using the line.

Appendix

SLUMP TEST

The slump test is used to measure the consistency of the concrete. The test is made by using a SLUMP CONE; the cone is made of No. 16 gage galvanized metal with the base 8 inches in diameter, the top 4 inches in diameter, and the height 12 inches. The base and the top are open and parallel to each other and at right angles to the axis of the cone. A tamping rod 5/8 inch in diameter and 24 inches long is also needed. The tamping rod should be smooth and bullet pointed (not a piece of rebar).

Samples of concrete for test specimens should be taken at the mixer or, in the case of ready-mixed concrete, from the transportation vehicle during discharge. The sample of concrete from which test specimens are made will be representative of the entire batch. Such samples should be obtained by repeatedly passing a scoop or pail through the discharging stream of concrete, starting the sampling operation at the beginning of discharge, and repeating the operation until the entire batch is discharged. The sample being obtained should be transported to the testing site. To counteract segregation, the concrete should be mixed with a shovel until the concrete is uniform in appearance. The location in the work of the batch of concrete being sampled should be noted for future reference. In the case of paving concrete, samples may be taken from the batch immediately after depositing on the subgrade. At least five samples should be taken from different portions of the pile and these samples should be thoroughly mixed to form the test specimen.

The cone should be dampened and placed on a flat, moist nonabsorbent surface. From the sample of concrete obtained, the cone should immediately be filled in three layers, each approximately one-third the volume of the cone. In placing each scoopful of concrete the scoop should be moved around the top edge of the cone as the concrete slides from it, in order to ensure symmetrical distribution of concrete within the cone. Each layer should be RODDED IN with 25 strokes. The strokes should be distributed uniformly over the cross section of the cone and should penetrate into the underlying layer. The bottom layer should be rodded throughout its depth.

When the cone has been filed to a little more than full, strike off the excess concrete, flush with the top, with a straightedge. The cone should be immediately removed from the concrete by raising it carefully in a vertical direction. The slump should then be measured to the center of the slump immediately by determining the difference between the height of the cone and the height at the vertical axis of the specimen as shown in figure A-1.

The consistency should be recorded in terms of inches of subsidence of the specimen during

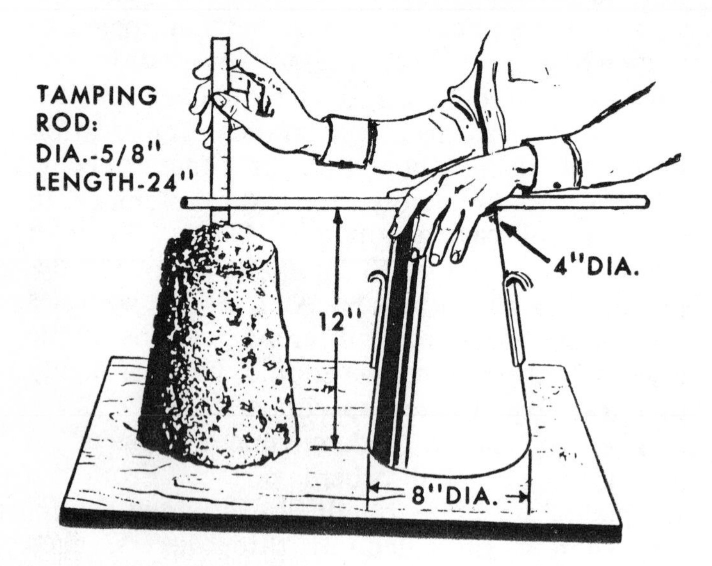

Figure A–1. —Measurement of slumps.

the test, which is called slump.. Slump equals 12 inches of height after subsidence.

After the slump measurement is completed, the side of the mix should be tapped gently with the tamping rod. The behavior of the concrete under this treatment is a valuable indication of the cohesiveness, workability, and placeability of the mix. A well-proportioned workable mix will gradually slump to lower elevations and retain its original identity, while a poor mix will crumble, segregate, and fall apart.

Index

R

S